1984

TELECOMMUNICATIONS

ANNENBERG/LONGMAN COMMUNICATION BOOKS
George Gerbner and Marsha Siefert, Editors
The Annenberg School of Communications
University of Pennsylvania, Philadelphia

Jerry L. Salvaggio
University of Houston

TELECOMMUNICATIONS

Issues and Choices
for Society

Longman
New York & London

Telecommunications
Issues and Choices for Society

Longman Inc., 1560 Broadway, New York, N.Y. 10036
Associated companies, branches, and representatives
throughout the world.

Developmental Editor: Gordon T. R. Anderson
Editorial and Design Supervisor: Diane Perlmuth
Manufacturing Supervisor: Marion Hess
Production Supervisor: Ferne Y. Kawahara

Library of Congress Cataloging in Publication Data
Main entry under title:

Telecommunications, issues and choices for society.

  (Annenberg/Longman communication books)
  Bibliography: p.
  Includes index.
  1. Telecommunication—Social aspects—Addresses,
essays, lectures. 2. Telecommunication policy—
Addresses, essays, lectures. I. Salvaggio, Jerry Lee.
II. Series
HE7631.T43 1983   302.2   82–17079
ISBN 0–582–29011–2

Manufactured in the United States of America

For my parents and Lorraine

Acknowledgments

This book was conceived simultaneously with a symposium which I organized on a similar subject in 1981. I am thus indebted to those individuals and organizations which provided support for my research on the Information Society and for the 1981 Edward R. Murrow Symposium. Thomas L. Kennedy gave me moral support and encouragement. Several organizations, including General Telephone, Northwest Pacific Telephone and the Haas Foundation provided generous financial support. Susan Tretivikk provided research assistance beyond the call of a teaching assistant.

I am also indebted to those individuals who provided insightful comments after reading earlier versions of the book. Richard Nelson, Garth Jowett, and Jean Hlavaty of the University of Houston were especially helpful in this regard.

For permission to use Daniel Bell's essay I am grateful to the Harvard Business Review. I am also grateful to Anthony Oettinger of Harvard's Program on Information Resources Policy for permission to use Bill Read's essay.

I should also like to thank George Gerbner and Marsha Siefert, editors of the Annenberg/Longman series for their wise counsel on editorial matters. Gordon Anderson and Diane Perlmuth of Longman are also to be thanked for providing valuable suggestions and editorial expertise.

Finally, I want to extend my warm personal appreciation to my wife for longstanding support and her excellent proofreading ability.

Contents

PART 2 PUBLIC POLICY FOR AN INFORMATION SOCIETY

Foreword

When the future of telecommunications is talked about, one is likely to find that the discussion focuses on one of two areas. The first centers on the marvels of telecommunications and the second on the impact technology will have on programming in the information society.

Technology

The traditional world of broadcasting is undergoing momentous changes as a result of the proliferation of new hardware. Indeed, the effects will be revolutionary, but most people are not really aware of them because of the evolving nature of these changes. Cable came on the scene largely to augment the broadcast signal and to reach remote areas. Now it not only can perform those tasks but also can immensely enlarge broadcasting's programming potential, both through an increased use of the satellite and a greatly increased number of channels. Videotapes can bring directly into the home a capability to choose the time at which one will see programs. Computer tie-ins enhance the usefulness of broadcast materials by making it possible to stop the show and study individual frames, or to engage in two-way communication. Low-power television stations offer an economical way to bring diversity to the home or school and to reach groups too small to attract access time in traditional ways. Direct home broadcast via satellite for specialized kinds of programming can be received by small, rooftop dishes and controlled by decoding equipment.

Videodiscs, videocassettes, direct satellite-to-home broadcasting, pay cable, teletext, multipoint distribution service, and fiber optics illustrate the bewildering world of telecommunications that could make over-the-air broadcasting as dated and obsolete as the crystal radio set of my boyhood.

The explosive proliferation of these and other new means of delivering entertainment, instructional and informational programming, as well as a host of other services such as electronic banking and shopping, interactive video, electronic newspapers, and home and business security monitoring, has created a communication of abundance unlike anything dreamed of by Jules Verne.

But, like the specter of "Big Brother" in George Orwell's prophetic *1984*, the emerging technologies that will form the base of the information society have raised serious questions about their implications for personal liberty, the increased possibility of a centralization of power, illegal surveillance, access to information, competition, and the proper role of government regulation. Clearly, telecommunications will alter the way in which we conduct our personal and national lives.

It has been said that communication is the glue that holds society together. Today the world depends on communication technologies—television, satellites, laser-transported messages—that did not exist a generation ago. But, all too often, an emphasis on technology seems to take precedence over concerns about quality, substance, and content. Burgeoning developments in the hardware of delivery systems have outpaced the formulation of public policy and an understanding of the problems in society that those systems might create.

Programming

Aside from technology, another aspect of telecommunications is enormously important, namely, the software. Discussions revolve around the implications of programming and are reflected in many different ways.

In the informational aspect, telecommunications meetings have been dominated by questions having to do with freedom of the airways and have featured several different points of view. The Western world has been understandably concerned about government domination of the airways. The Iron Curtain countries have been equally concerned that their governments might be subverted by ideas not compatible with their own ideology. Third World countries, sensing themselves dominated by powerful forces in each of the other camps, feel restricted in developing an approach of their own. This is a battle that will continue with perhaps ever increasing intensity, and its outcome may have a good deal to say about the information society.

From the national perspective, in America we continue to debate whether the violence that now appears on television affects people's behavior or relates to the fierce controversy over gun control. And when we are not worrying about the impact of violence, we wonder whether the dominance of sex on television is contributing to the breakdown of our values and the family structure. Hovering over these debates is the First Amendment to the Constitution, and how we can square our moral and ethical concerns over programming with the concept of free speech.

In the world of education, there are debates over whether the much-admired *"Sesame Street"* series educates and fascinates children or whether it shortens their span of interest for the "real" education that will come into play when they are old enough to go to school.

The remarkable innovations in the technologies of delivering programs and services do not necessarily mean improvements in the quality of programs or services. Do more program choices make for a more enlightened public? Does increased access to information make for a more informed citizenry? Has the technology of delivery outstripped our ability to process and make use of the information delivered? Where will the programming needed to fill the new channels of communication come from, and who is going to pay for it?

In Salvaggio's book a number of thoughtful researchers, scholars, and observers address various implications of evolving telecommunication systems. The chapters focus on critical issues: the nature of an information society, the First Amendment and the information society, the social impact of telecommunications, life in the electronic future, monopoly versus competition, and the social effects of media convergence. One observer, Joseph N. Pelton, notes that "modern telecommunications [technologies] . . . bring with them enormous political, economic, and social problems. Yet, one cannot hide the fact that they also bring with them enormous opportunities to diminish illiteracy and even opportunities to aid the cause of world peace."

Suggestions on how we might confront and resolve our problems and capitalize on our opportunities in large measure are likely to shape and define information societies in the twenty-first century.

Robben W. Fleming

Preface

Dimensions of the Information Society

The telecommunications revolution this book discusses is a consequence of the rapid convergence of "computer and communication technologies." This combination allows us to acquire, store, manipulate, and transmit information at rates and in ways hardly imagined a generation ago.

To help understand the broad social dimensions of the information society, we might compare present and emerging communication capabilities with an earlier revolution—changes in transportation. Mobility in physical space is achieved through roads and railways; mobility of information is gained through the telecommunications network. Telephone cables, microwave systems, and satellites are the transmission analogs to the transportation grids and modes of the industrial economy. But where expansion in transportation capabilities and, more generally, increasing industrialization has, since the early nineteenth century, required increasing energy and material resources, the opposite trend characterizes the information society. Human knowledge is its primary resource need. This is its principal claim to being a unique revolution and the basis for the claim that it is likely to result in social and economic impacts that at present can be only dimly preceived.

Present and Anticipated Technical Capabilities

During the last half of the 1970s the number of components that could be put on a single computer chip increased a hundred-fold. This rate of progress is likely to continue and to result in a ten-thousand-fold increase in performance at roughly the same cost.

Computer circuitry cost approximately $1 million in 1955. Before the end of the century, one 1955 penny may well purchase a device a thousand times more powerful than the original million-dollar computer.

Because of developments in Very Large-Scale Integration (VLSI), million-bit computer chips or even a complete computer on a wafer are possible. The functions of microprocessors, which are now postage-stamp size, will likely be performed in the future by a device that is the size of a grain of salt.

At the same time, progress in communication technology is dramatically increasing the volume of information that can be transmitted simultaneously. Advances in satellite technology not only make the international transfer of information much easier but also improve the quality of transmission, which is increasingly insensitive to distance. But although satellites have already done more to unite the world's communication than any previous technology, their potential has only begun to be tapped.

In an information society communication and computing functions merge into a combined system. Typewriters and computer terminals became "intelligent." Telephones not only can handle voice messages but also can process and manipulate information in whatever manner the user desires. Home data processing, computer-linked people and services, and a total information society are likely results.

Institutional Aspects

The information society has a number of important institutional aspects that are worth noting. First, its scientific and technical foundations were laid by unusually fruitful, long-range cooperation between university and industrial laboratories. Second, a wide range of scientific and engineering disciplines contributed, and continue to contribute, to these foundations. Emerging information-handling capabilities have, in turn, stimulated new, fruitful research directions in several disciplines and have themselves contributed to further progress. Third, as the expanding information revolution provides new capabilities in nonscientific and nontechnical areas, professionals in these areas have become stakeholders in the further development—and perhaps control—in the directions the information society is taking.

Impacts on Selected Areas

As Marc Porat has pointed out, more people in the United States are involved in handling information than are employed in mining, agriculture, and manufacturing; and the importance of the information sector of the U.S. economy is certain to increase. Computer technology will be at the center of this activity. Conservative estimates attribute at least 15 percent of industrial productivity growth in the 1970s to the use of computers. In the telecommunications industry, between 1960 and 1977 productivity rose 5.5 percent annually, almost triple the rate of the nonfarm business economy. White-collar jobs and service industries are likely to experience the greatest future productivity increases. The present outpouring of products for the office of the future promises to improve office performance substantially.

Some effects of the information society on the conduct of science and on education are worth noting: The impact of the telecommunication revolution on the conduct of science is best exemplified by the so-called smart instruments appearing in research laboratories—instruments possessing amazing sensitivity as a result of their ability to process large quantities of data. Similarly, miniaturization is having an important effect on the scientific measurements that may be possible. For example, future biological research will be influenced by the capability to implant microinstruments permanently in biological structures. The possibility of seeding atmospheric and oceanographic experiments with cheap and disposable sensing and data processing devices will also lead to new ways of doing science.

Computer-based information and retrieval systems promise to put an automated library and a reference librarian in the office of every scientist. We are already seeing this in interactive cable systems. Virtually all participants in the scientific enterprise will be affected by such retrieval systems—the scientist who writes articles, editors, referees, publishers, librarians, and users of the information.

The information society will clearly have a tremendous impact on education. The widespread availability of computer terminals, for instance, could permit a breakthrough in education on the scale of the printed book. Most of the subjects traditionally taught in the schools could be available in the future on computers, and in a form that would make them highly adaptive to each student's individual interests, abilities, and needs. Such individualized instruction, available to everyone when and where they want it, would certainly change society's view of education.

Policy Issues, Ethical Issues, and Value Conflicts

Rapidly emerging capabilities inevitably challenge traditional ways in which things are done—in the office, the factory, the home, the laboratory, and the classroom. Thus value conflicts and ethical problems are already surfacing in the information society.

Regulatory Questions

Many questions concerning regulation of information must be re-examined. Traditionally, the federal government has regulated telecommunications. Should it, then, regulate information disseminated electronically, and thus perhaps electronic sources of that information as well? How can regulations be framed so that public access to information is maximized while proprietary rights of producers and developers of information are protected? With respect to data bases, what constitutes the public domain? How can we protect intellectual property and appropriately reward its producers while preventing the kind of piracy that has become commonplace in the music industry?

A significant shift from the use of hand-carried mail to electronic mail will soon become feasible. If electronic message systems (EMS) are considered a communication service, then, according to the law in 1982, they must be regulated. Does it follow that the U.S. Postal Service should be regulated? If the Postal Service is permitted to offer EMS at prices below cost, what will be the effects on private industry?

Privacy Questions

How can personal privacy be maintained given the virtually unlimited storage capacity of data banks? Can means be devised to assure the privacy of individual records while making aggregated statistical information available for decision-making purposes? (Examples are aggregated medical data required for longitudinal epidemiological studies; longitudinal achievement data required for evaluating educational programs.) How can the security of data and program integrity be protected against penetration by outsiders, who may operate with great skill from remote terminals?

International Issues

Many nations do not view expanded satellite communication and re
lated satellite remote-sensing capabilities as entirely beneficial. Rather,
as Herbert Schiller reminds us in Chapter 2 of this book, they see these
developments as threats to their sovereignty. What international institutions can assure unimpeded transborder information flow while
protecting legitimate sovereign rights?

The use of communication satellites is expanding so rapidly that
several less developed countries fear that available geosynchronous
locations and usable portions of the electromagnetic spectrum will be
saturated before they are able to implement systems for their own purposes. How can the orderly, international development of communication capabilities be assured?

Social Control of Science and Technology?

The foundations of the information society and its profound social and
economic consequences derive from a fruitful series of advances in
scientific and engineering disciplines. These advances were possible,
in large measure, because of the extraordinary individual and institutional scientific freedoms enjoyed in the United States: the freedom
of individual scientists to pursue promising research directions
wherever they led; the freedom of institutions to exploit the results
of their investments in scientific and engineering capabilities.

But science and technology must ultimately be consistent with the
public good; these freedoms are not unlimited. As with the information revolution, when advances in science and technology lead to difficult policy, social, and ethical questions, what, if any, are the special
responsibilities of scientists and engineers?

Professional and Social Responsibility

Most people would agree about the dimensions of the professional responsibilities of scientists: to pursue excellence, to remain as objective
as possible in drawing conclusions from data, to share research results
and procedures with colleagues in open literature, and to resolve dis-
agreements about conflicting scientific theories on the basis of scientific
evidence. Whether scientists and engineers in an information society
have social responsibilities that go beyond the responsibilities of all
citizens is not an easy question to answer. Various authors in this book
attempt to provide insight into the complexity of the information
society.

In many cases, scientists and engineers may have earlier and deeper insights than nonscientists into the possible social impact of their work simply because they are more intimately and deeply involved with laying the foundations for social change than nonscientists are. It can be argued that in such cases scientists have a responsibility to assist the public to understand the complexities of scientific research and the uncertainties in estimating its risks and benefits. Scientists have a responsibility to try to deepen their own understanding of the ethical and social implications of their work. And scientists must be prepared to exercise self-regulation of research activities when appropriate, or to join with nonscientists in monitoring external controls if necessary.

Historical Examples

The social responsibility question, while a difficult one, has been with us from the beginning, as some historical examples suggest. The trial of Galileo in 1635 remains a classic case study in the value conflicts that can result from advances in scientific understanding. In 1660 the group of English natural philosophers who banded together to form the Royal Society of London, perhaps mindful of Galileo's example, found it advantageous to seek a royal charter granting freedom to pursue their inquiries, provided they did not touch upon matters affecting church and state. In so doing they agreed to regulate their own activities, to assure themselves the maximum possible freedom consistent with the prevailing social values.

Following the discovery of nuclear fission in 1938 and prior to America's entry into World War II, a group of U.S. physicists and their refugee colleagues from Europe decided, for national security reasons, not to publish any research results on the fission of uranium in the open literature. Their action was completely voluntary and took place before the U.S. government was aware that nuclear physics might have military significance. After the war, when the immediate threat to national security had passed, many of these physicists took the lead in insisting on a full and informed debate about the use of nuclear power and the proliferation of nuclear capacity. Their efforts were instrumental in helping to achieve several important objectives including civilian control of atomic energy and, on the international level, the 1962 treaty banning atmospheric tests of nuclear devices.

Molecular biology is the field that has experienced the most rapid and dramatic advances and has been an area of well-publicized conflict between scientific and social values. Recombinant DNA techniques, developed less than a decade ago, provide biologists with a powerful research tool for studying the structure of DNA molecules, which are the basis of heredity. Since these molecules contain the genetic information required by nature to fabricate proteins, recombinant DNA techniques also provide the basis for cloning a wide range of biological substances. Thus they have significant implications for medicine and agriculture. Indeed, several firms have been established to explore the commerical applications of recombinant DNA in the production of pharmaceuticals, and in the summer of 1980, in England, insulin produced by means of the technique was tested on human subjects.

These rapid advances in molecular biology offer unprecedented capabilities for understanding and manipulating the genetic information that constitutes the basis of life itself. As such, the current revolution in molecular biology can be characterized as an information revolution on a more precise and fundamental scale than is being considered in the essays that follow.

In 1973 several biologists who had pioneered the recombinant DNA techniques recognized that whereas the revolution it promised could convey enormous social benefits, it could also have less desirable consequences. The immediate concern of these biologists was that new and potentially dangerous biological forms might inadvertently be cloned. Mindful, perhaps, of the earlier initiative of the nuclear physicists, they called for a worldwide moratorium on recombinant DNA research until the risks of this new technique could be assessed.

This group of biologists took seriously their responsibility to convey their misgivings to their colleagues in other scientific fields, to the press, and to federal agencies (primarily the National Science Foundation and the National Institute of Health) that supported recombinant DNA research. Their intitiative had several results. First, it led to the development of guidelines and procedures by NIH for pursuing federally sponsored research so that research would result in the minimum possible risk to the public. Second, it engendered research aimed at determining the probable extent of alleged risks. Finally, there was considerable public controversy in several communities about whether the NIH guidelines were, in fact, adequate to protect the public's safety. Results of research during the late 1970s indicate that most of the original fears about recombinant DNA were groundless, and also suggest that the technique may offer even more benefits than originally envisioned. As a result, public fears have largely abated.

While many molecular biologists recall the recombinant DNA controversy of the 1970s as a traumatic period, others believe that nothing was lost in the long run and that a good deal in the way of public confidence may have been gained. At the least, many more people are aware of the fact that we are on the threshold of a new revolution in biology than would otherwise have been the case.

Broad Lessons

As these examples illustrate, there are occasions when the professional and social responsibilities of scientists may conflict with each other. For example, the professional responsibility to publish openly or to pursue research directions wherever they may lead may conflict with a broader perceived responsibility to the public good. In those cases, scientists may decide to limit their scientific activities.

Individual scientists and engineers have often chosen to exercise what they regard as their social responsibility by involving themselves in public debate about policy issues with scientific content. Since the social awareness and political persuasions of scientists exhibit the same diversity as found among any group of professionals, their responses to the social responsibility question have also been diverse. This has led to some public confusion about the technical basis of such policy issues. But despite the not-surprising fact that scientists differ in their value judgments, public debate on these issues cannot help but be better informed when scientific and technical components are aired.

The obvious conclusion that can be drawn from history is that the social responsibility question will certainly continue to be important in the future as the results of science and technology become ever more pervasive. This has particularly important implications for education. Since scientists and engineers are bound to be faced with difficult value conflicts during their professional careers, their education must prepare them for the fact that conflicts are likely to arise and equip them to cope with those conflicts. Equally important—perhaps more important—education must equip the nonscientific public to understand and cope with a world that will continue to change rapidly as a result of science and technology. At a minimum, nonscientists need to understand both the power and limits of science and technology and their multiple relationships with social, ethical, and policy issues. In this area, we need to do a much better job than we have to date.

Conclusions

Unlike innovations derived from advances in nuclear physics or molecular biology where the social and economic impacts are accompanied by potential physical catastrophes, the communications revolution is likely to have a limited physical impact. We can anticipate many benefits, however, assuming that we make the optimum use of the capabilities of computer/communication technologies for the public good. Determining information technologies that will be in the public good will requires a great deal of wisdom.

Scientists and engineers can contribute their own special insights—born of their training and experience—to public discussions about the best social uses of information and communication technologies. They can and should also contribute to public discussions about the whole range of science- and technology-based innovations. The best way for scientists and scientific institutions to continue to enjoy the freedoms that have been so central to the scientific enterprise and to our society may be to demonstrate that scientists and engineers understand that those freedoms entail the acceptance of a measure of social responsibility.

The essays in this book demonstrate their deep concern for social responsibility. Their concerns run the gamut from invasion of privacy to information control. They point out the most salient issues and offer suggestions for developing communication technologies without adverse social effects. Many of these issues are central to the development of public policy in an information society.

John Slaughter

About the Authors

Bell, Daniel, Ph.D., Harvard University. Daniel Bell is Henry Ford II Professor of Social Sciences at Harvard University. He is the author of numerous books including *The Coming of Post-Industrial Society* and *The Cultural Contradictions of Capitalism*.

Homet, Roland S., Jr. Roland Homet is a communications lawyer specializing in strategic counseling and was formerly Director of International Communications Policy at the U.S. International Communications Agency. He has been Director of the Aspen Institute Program on Communications and Society and has recently published *Politics, Cultures and Communication*.

Ogasawara, Ryuzo. Ryuzo Ogasawara, a graduate of Tokyo University, is an executive staff writer for the *Asahi Shimbun*, Japan's most influential newspaper. In addition to several articles on the implementation of new communication media, he has published *Between Newspaper and Broadcasting*.

Pelton, Joseph N., Ph.D., Georgetown University. Joseph Pelton is the Executive Assistant to the Director General, Intelsat, the organization that operates the world's global satellite system, which interconnects 150 countries, territories, and possessions. He has authored three books and a number of articles and reviews on communication technology. His most recent book, *Global Talk*, was nominated for the Pulitzer prize.

Read, William H. William Read's career in communication includes work as a lawyer, journalist, and scholar. While a faculty fellow at Harvard, he wrote numerous articles and several books on communication, including the critically acclaimed *America's Mass Media Merchants*. Read is Communication's Counsel with the U.S. International Communications Agency.

Rimmer, Tony, Ph.D., University of Texas. Tony Rimmer's research is in the area of communication technology. He teaches Journalism at Indiana University.

Rose, Ernest D., Ph. D., Stanford University. Ernest Rose has been a documentary film producer for thirty years. He is past Dean and present Professor of Communications of the Liberal Arts College at California State Polytechnic University, Pomona. Rose has contributed to six books and has written in more than thirty professional journals.

Salvaggio, Jerry L., Ph.D., The University of Michigan. Jerry Salvaggio teaches communication technology at the University of Houston. He is the author of several articles and book chapters on various aspects of telecommunications. His primary research interest is the social impact of communication technology.

Schiller, Herbert I., Ph.D. Herbert Schiller teaches communications at Third College, University of California, San Diego. He is the author of numerous articles and several influential books on communications. His most recent book is *Who Knows: Information in the Age of the Fortune 500*, published by Ablex Publishing Co.

Slack, Jennifer Daryl, Ph.D., University of Illinois. Jennifer Slack's speciality is the relationship between communication technologies and sociocultural structures and values. She teaches communications at Purdue University. Slack is the author of *Communication Technologies and Society: Conceptions of Causality and the Politics of Technological Intervention*. With Fred Fejes, she is also editing *The Ideology of the Information Age*.

Slaughter, John, Ph.D., University of California, San Diego. John Slaughter is the author of numerous scientific studies. He is past Vice President and Provost of Washington State University and is past Director of the National Science Foundation. Presently Dr. Slaughter is Chancellor of the University of Maryland.

Introduction

Exactly what is an information society? One scholar submits that the phrase is the sort of trendy label that satisfies our penchant for slogans. Another scholar begins his book on the subject by merely noting that the phrase is troublesome. Marc Uri Porat offers a precise "economic definition" of information societies. For Porat, at least half of a society's labor force must be engaged in information-related work for that society to be an information society. According to Porat's criterion, the United States, Japan, Sweden, England, and a number of other countries are information societies. Yet this tells us little about the infrastructure of information societies, the potential of telecommunication systems, and what the information society portends for the future.

A brief historical survey of the literature on telecommunication systems and the notion of future global networks might provide some insight into what is meant by an information society. A catchy phrase that served as a predecessor of the term information society was Marshall McLuhan's "global village." McLuhan, before anyone else, used the term to describe the way in which the electronic media would homogenize the various societies of the world into a common social network. The electronic age, he argued, was similar to the character of our central nervous system. Our central nervous system, McLuhan advised, is not merely an electric network but constitutes a single unified field of experience.

Almost simultaneous with McLuhan's rhetoric came optimistic predictions of a wired world. In 1971, *U.S. News and World Reports* published a book entitled *Wiring the World: The Explosion in Communications*. Said the book's author: "We will become a wired nation, pioneer of a wired world. Over-the-air television broadcasting would be supplanted in large part by 'broadband' networks bringing unlimited programming as well as two-way communication into home and offices."

Other Delphic oracles were equally disposed to take a favorable view when reflecting on the consequences of global telecommunication systems. Dieter Kimbel from the Organization for Economic Cooperation and Development in Paris predicted a total decentralization—the disappearance of cities that McLuhan proposed. Mark Hinshaw drew two scenarios that had an eventual "complete" wiring of the nation in an orderly incorporation of the new technologies, rather than a patchwork wiring. Herman Kahn similarly projected a global metropolis. Wilbur Schramm predicted a world connected by closed-circuit television, and Arthur Clarke foresaw the demise of cities as we know them.

The authors in this book use the term "information society" rather than wired world or global village. They are far less optimistic than the earlier writers, but the society they describe would seem to be consistent with the above in that they refer to the future. The common denominator for the numerous phrases coined to describe societies of the future is the emphasis on telecommunication systems and the easy access to information. Thus, Yoneji Masuda, in *The Information Society as Post-Industrial Society*, refers to "information utility" as the societal symbol of the information society. In such a society, Masuda observes, "anyone, anywhere, at any time will be able easily, quickly, and inexpensively to get any information which one wants to get." This may or may not be the case in most information societies, but it would seem to be what scholars and futurists have in mind when they refer to the information society. Information society, then, alludes to what awaits us in the future. It is a society in which individuals transmit and receive news, data, and entertainment through home telecommunication centers. In the information society, information is a valued commodity

produced, collected, stored, distributed, managed, traded, bought, and sold like other commodities.

If "information society" is without a precise definition, the term "telecommunication" is used in this book to denote communication systems ranging from standard broadcasting and cablecasting to satellite and computer communication systems. While some scholars continue to distinguish between computer and telecommunication systems, the two are referred to here simply as telecommunications, for the systems are quickly becoming inseparable.

Exactly how extensive and interconnected telecommunication systems need to be for a society to be considered an information society remains vague. The authors in this book, though, consistently refer to the marriage of computers and communication systems. The interface of these two systems is what makes national information networks possible. Another element the contributors direct attention to is the use of telecommunication systems in the home as well as at work. In an information society, telebanking, telecourses, televoting, telemetering, telecommuting, telegames, and teleshopping are common. Put another way, members of an information society are tied into a vast electronic telecommunications network.

This is the information society that has scholars and laypersons alike rereading Aldous Huxley and George Orwell and forewarning us to greet the information society with caution and a degree of skepticism. Such apprehension, as represented in the essays in this book, is based less on paranoia than on the universally recognized fact that telecommunication systems are growing at a pace that far exceeds our ability to divine the social consequences or develop a sound public policy. Hardware systems alone have escalated at a pace that makes it impossible to forecast with any accuracy their cultural-social implications. In addition to cable television, communication satellites, fiber optic networks, and personal computers, other forms of telecommunications have been in vented, tested, and put into operation before scholars could even begin to develop methodologies for assessing their cultural and social effects.

The following telecommunication systems and information services are but a fraction of the telecommunication systems that will be widely used in information societies: personal computer terminals, handheld computers, computer conferencing, computer databases, electronic mail, electronic polling, electronic voting, electronic bulletin boards, electronic voice systems activating videodisc storage systems, home facsimile terminals tied into direct broadcast satellites, interactive cable TV, specialized cable networks, gas plasma large-screen TV sets, telebanking, teleshopping, telecourses, laser copying systems, teletex, viewdata, holographic storage, video conferencing, videocassette recorders tied into home computer terminals and videodiscs, video newspapers, and video magazines. If most of these systems' hardware and software were developed and introduced into society one at a time and over a thirty-year period, one might not be so concerned about the consequences, for they would be easier to predict. Yet these systems are being introduced not only in an explosive fashion but, in most cases, similar though incompatible telecommunication systems are being introduced simultaneously. To gauge the magnitude of the problem of determining what the eventualities could be, one need only recall the thousands of scientific studies done on the social effects of one telecommunication system—the traditional TV—and how much there is still to learn.

What Eric Fromm said about Orwell's *1984* might be an apt summary of the essays in this book: They are an expression of a mood and a warning. To what extent the authors believe the information society will be an incubus descending on the twenty-first century is not certain. Nevertheless, each author questions the impact the telecommunications revolution will have on an individual's liberty, right to privacy, and equal economic opportunity. Three contributors caution that computers belonging to cable companies, corporations, and the government will be used for monitoring the purchasing and viewing habits of individuals. Daniel Bell discloses the case of a physiologist who discovered a device that could measure an animal's heartbeat, respiration, muscle tension, and body movements from a distance. The physiologist was unaware that what he had invented was the same device that Orwell had written about in *1984*. On further investigation, the physiologist found that 100 of the 137 devices imagined by Orwell were already practical. Bell's point in recounting the experience of this physiologist is that knowledge and science are progressing at a pace even scientists cannot comprehend. Bell's essay is especially effective in providing a sociological study that places the information society in historical perspective.

Herbert Schiller articulates more emphatically than the other authors that telecommunications is likely to be used primarily for surveillance, monitoring, and marketing. He argues vigorously that the military-originated information technology, and the military will continue to dominate the information field for its own purposes.

Ernest Rose's essay lays special stress on the advantage telecommunication gives to big business as opposed to the individual. At the same time he reminds us of how easily computer thieves stole $21.3 million

from a Wells Fargo bank. Joseph Pelton is more positive than some of the authors, noting that the range of services available to the home consumer will be broader and will cost less in an information society. On the other hand, he asserts that we may suffer from a loss of individual liberty and could become the victims of terrorists who gain access to information systems and of thieves with knowledge of the technology.

Ryuzo Ogasawara and Jerry Salvaggio discuss the situation in Japan where the notion of becoming an information society has been on the minds of scholars, government officials, and corporate executives since 1965. Telecommunications in Japan have not been developed on an ad hoc basis but with the information society as an objective. Yet even this kind of preplanning may turn out to be a two-edged sword if the public is complacent and allows the government and the information industry to do all the preplanning.

The essays in Part 2 discuss the fact that the United States at present has no comprehensive public policy for the introduction of telecommunications on a national scale. Few objectives have evolved, and little or no action has been forthcoming in the area of regulatory changes as the government maintains an incurious and apathetic attitude toward the notion of becoming an information society. The authors make the point that major policy issues need to evolve and be carefully conceived if we are to minimize the consequences. Roland Homet, an exponent of marketplace competition, questions the extent to which telecommunications should be regulated. He argues that existing laws, such as those dealing with copyright, banking, and privacy protection, are sufficient and that the government should not intercede with economic regulations.

A related problem, addressed by both William Read and Tony Rimmer, is that existing regulations are woefully inadequate in view of the manner in which telecommunications is modifying the traditional ways news is received. They seem to agree that videotex and direct broadcast satellite have blurred the distinction between print and broadcast journalism. This, they point out, raises a particularly troublesome question, which Congress will need to address. Should electronic information systems be regulated according to the First Amendment or according to existing FCC regulations?

If the information society will require major policy decisions, how should the U.S. proceed? The FCC thus far seems to be disinterested. And, even if the FCC were less neutral, most people believe it is incapable of doing the research required. Nor does the FCC have the financial resources required as a basis for developing experimental systems

or a long-range public policy. Jennifer Slack's article is directed at our failure to correctly assess technology. She makes a strong case for her argument that technology assessment, especially in the area of tele-communication, is not up to the task demanded by the information society.

The concluding chapter surveys the negative literature on the in-formation society. Information control and the manipulation of in-formation technology are cited as recurring themes in the essays in this book. If the information society is not to be the nightmare many are forecasting, then the public will need to be less apathetic, more skep-tical, and more knowledgeable than it is today about telecommuni-cation systems and the problems that may be inherent in an informa-tion society.

In addition to the issues discussed in this book, there are a plethora of questions to be answered in the international arena. Schiller, Pelton, Homet, Salvaggio, and Read have examined these issues in other books and articles, which are listed in the bibliography. This book attempts only to present essays that examine potential side effects associated with information societies in general and to identify major telecom-munications policy issues that need to be addressed in the United States if adverse effects are to be minimized.

That there will be adverse effects is beyond question. Rose's com-ments succinctly sum up the feelings of the other authors in this book when he writes: "Even a casual glance at history reveals that major changes in technology have always brought with them choices with which individuals and societies have been ill prepared to cope. Small wonder that the specter of an upheaval as potentially radical and pervasive as the approaching 'information society' carries with it high levels of anxiety along with great hope and promise." Whether the side effects will be closer to a nightmare than a utopia will depend on the extent to which we are prepared for the challenge. Becoming knowledgeable of what lies ahead is the first step toward meeting that challenge.

Jerry L. Salvaggio

PART **1**

THE NATURE OF
INFORMATION
SOCIETIES

Chapter One

Moral and Ethical Dilemmas Inherent in an Information Society

Ernest D. Rose

Even a casual glance at history reveals that major changes in technology have always brought with them choices with which individuals and societies have been ill prepared to cope. Small wonder that the specter of an upheaval as potentially radical and pervasive as the "information society" carries with it high levels of anxiety along with great hope and promise.

In thinking about the scope of telecommunications and its ramifications for information societies, I am reminded of a lighthearted article that appeared in *New York* magazine in 1976. It was written by Jon Bradshaw and entitled "The Shape of Media Things to Come." Speaking from some undisclosed point in the future, the author looks back on the 1970s and '80s, viewing with amusement the primitive state of human relations and the role that communication technology had in shaping that curious society. Because it may just possibly harbor some grains of truth about future moral and ethical dimensions of tomorrow's mediated world, I have chosen to include extended passages from these futurist speculations here.

> ... in retrospect, man's lack of ingenuity was almost unbelievable. ... It was
> not until copper wires had been replaced with broad-band optical fibers that

communications became somewhat more refined.... [It] meant that users could link up with central information services in order to hear what they wished to hear instead of what someone else wanted them to hear.... All shopping was performed through the computer-telex and all mail and newspapers were now effected through computer-television.

The large TV screen, modern so short a time before, was replaced with a three-dimensional wraparound wall screen. Even middle-class homes were glutted with voice-activated typewriters, picture phones, hologram-projection machines, and laser burglar alarms set not to kill but to stun....

The home had now become a total environment—the ultimate cocoon—and life, its bemused inhabitants believed, was terribly modern. More important, and for the first time, an intimate dialogue had been initiated between man and machine. The computer had eliminated the merely mechanical. All essentially mechanical movements, in fact, were becoming obsolete, although they had not yet been taken ... to their purely logical conclusions.

Almost no one "traveled" anymore. More and more business was conducted from the home. Students were educated in the home; and as an alternative to ... enforced holidays, the three-dimensional wraparound screen allowed one to *be* in say, the South of France, and the sound-around system made soft Mediterranean noises....

The computer, or Central Processing Unit as it came to be known, controlled everything. The demand for communication between computers was filled by the laser. The portable C.P.U., worn as a watch, was a peripheral device linked by matching light frequencies to the main C.P.U.... The tiny device functioned as a calculator, watch, telephone, and memory bank. You didn't have to know what you wanted from the C.P.U.; it calculated that for you. It always knew what you wanted because it knew you.

In the home itself, manual lighting had been replaced by glow walls, which stored light by day and emitted it by night. The kitchen had become redundant. In its place, each home was equipped with a photosynthesis module that provided highly enriched liquids for human nourishment....

The art world, too, had been revolutionized. Homes of the period were decorated with multiflex neon sculptures, frozen light murals, and dancing three-dimensional laser paintings....

No one actually wrote anymore. Books had long since disappeared.... Information was now being transmitted by light, almost always when one was asleep, so that on waking one had already [received the entire contents of] the "newspaper."

The computer had become self-evolving, self-sustaining. It performed its own maintenance as well as physical repair. Unlike man, it never required a doctor.... Computers were used to build computers, of course. Only another computer, in fact, could provide the degree of accuracy necessary to breed a further generation of computers....

Although world communications had evolved considerably, one major flaw remained which impeded progress for many years.... Man had always been beamed into his environment. It was not until man learned to beam the environment into *him* that real progress, as we understand it today, was possible. . . .

As everyone now knows, the evolutionary process began to accelerate. All essentially mechanical movements, whether those of machines or of man himself, became slowly obsolete and man, in time, arrived at a new definition of himself. . . .

There had been a steady erosion of purely human senses. The swift advances in computerized communications only hastened the inevitable. The sense of taste was the first sense to disappear, then smell, then touch, and almost immediately thereafter, hearing. Speech had always been a highly mechanical form of communication, tiresome and time-consuming. . . . I have always been surprised that the spoken word lasted as long as it did, that it had not disappeared with the introduction of the telepathic computer, which permitted one to "talk" in a more direct and efficient manner. Once the telepathic computer—worn as a watch and linked directly to the nervous system—became fashionable, speech was no longer necessary. The new language created instant communication, instant comprehension. . . .

Computers . . . revolutionized the home, of course. Spaces were equipped with ultrasonic mood systems, especially sensitized to their inhabitants, so that intimate communication was possible on every level without people having to move from one room to another.

New thresholds demanded new anatomies. The human body itself, long an awkward weight to cart around, had become little more than a vexatious bore. The crude mechanical functions for which it had been created— hunting, walking, eating, reproducing—were antiquated now. . . .

The last of the crude senses to atrophy was sight. Given what seems, even to such a cynic as myself, the really extraordinary developments in cosmic technology, it was soon no longer necessary to open the eyes. Why bother? Entire universes were available when the eyes were shut. It was the last of the old sensory barriers to be overcome.

It is difficult and not a little tedious to catalog the complicated advances that have taken place since then. . . . Suffice it to say that one occupies infinity. And I've not left this particular space for twenty years or more. There is no longer any "time," at least as you would comprehend it, so it is difficult to be precise. There are no "people" as such. There is no "sign of life." Consider it another way. As you must know, the source of all life is light, and we have simply, if I may use so complex a term, become light again. And it amuses me, you understand, to use this primitive form of communication in order to tell you so.[1]

While I have no desire to defend the credibility of this author's flights of fantasy, it is nonetheless intriguing to realize how many things that were spoken about in futuristic terms in 1976 have now become realities. Of course, like all tools, advances in technology can be used by people for progressive or regressive purposes. But one need not go as far into the future as Bradshaw's imagination carries us to see some implications of telecommunications that are deeply troubling. They range across the entire spectrum of issues from privacy to piracy, from property rights to human rights, and from free market values to the value of a press that is free.

I

We are told that in the information society various forms of telecommunications will allow us to transform our homes into command posts from which we will do our shopping, pay our bills, make our travel arrangements, do most of our work, and satisfy our fondest entertainment needs. But even those within the communication profession express concern over the price, in terms of personal privacy, at which this freedom will be purchased. Says TV executive Joel Chaseman, of the *Washington Post-Newsweek* conglomerate, "How tight a castle will your home become when the computers of cable company and charge card outfits speak to AT&T's computer and together discuss what you've watched, where you've gone, what you've read, the source and size of your income and what calls you have made?"[2] The computers of the information society will surely know all this and more, and they will tell what they know to those who can profit from that knowledge (and pay for it), unless someone decides which of our records may and may not be released for scrutiny.

Such concerns lie not just in the information society. In his book *Friendly Fascism*,[3] economist Bertrand Gross reports that in early 1980 detailed plans were made to register the country's young people for the draft without their knowledge through a system of "passive" registration. It would be done simply by compiling a computerized list of names and addresses of men of a given age and assembling other information from school records, the IRS, the Social Security System, and state drivers' license bureaus.

The Watergate hearings revealed how casually IRS scrutiny of financial records was used as a political tool for the harassment of individuals and corporations. Telecommunications promises to enhance the potential not only for dossier building but for simplifying retrieval from many different sources at an ever increasing pace.

Whether or not one contributes to unpopular political causes or fails to contribute to the right political fund drives is already available to someone seeking that information. The Census Bureau and other federal agencies routinely amass personal data of an economic, social, political, and environmental character. Much of it flows through narrow channels, not readily available for monitoring or public scrutiny. And although government policies on information usually specify clear limits by which it must abide in soliciting information, computer data are often shared or exchanged between friendly agencies and even private-sector institutions with common interests. In most cases, the information then becomes the property of the recordkeeper, and the individual loses the ability to exercise control over it. At that point, a person has almost no recourse against others obtaining the information and using it in ways that provide influence and control over the individual's life.

Asked about the potential invasion of privacy inherent in cable TV customer monitoring, a senior vice president of QUBE in Columbus, Ohio laughed and said, "If someone wanted to bug your house, the telephone line is already there."[4] Certainly an information age brings with it vastly more sophisticated tools for penetrating the already flimsy veil of personal privacy. In 1966 one expert on surveillance included among the methods already technically feasible at that time:

- The use of any wires in the "wired society" as listening devices, even when the receiver is on the hook or the dial set at "off"
- Tapping of computer tapes and computer communication lines
- Checking on human movement and activity through remote sensing devices
- "Mail cover" techniques, which not only record the name and address of senders and receivers but scan the contents of correspondence
- Sensing and reporting devices embodied in credit cards and automatic tags
- Transponders implanted in the brain of arrestees released on bail, criminals on parole or probation, or patients leaving mental hospitals[5]

If all these methods were practical in 1966, one can only guess at what steps have been taken since then, given the pace of telecommunications development in information societies. One thing is certain. The legal protections relating to personal information and rights of privacy have not kept pace with social and technological changes, nor can we expect them to do so in the future. This increasing gap between what we can do and what we should be doing with technology is the most compelling reason why the ethical and moral dimensions of the problem must be directly confronted prior to the arrival of the information society.

II

We hear of other marvels that the computer age has made possible, like the desktop machine that can read aloud to the blind from a printed page. Talking computers are already able to listen to customer inquiries and give up-to-the-minute quotations on any stock or bond transaction desired. Lawyers employ electronic scanners to conduct searches of civil and common-law precedents. At a normal fee ranging upward from $70 an hour, searches that used to cost from $100 to $500 are now possible at a time cost of only $10 to $20. IBM's 3850 computer library can store the text of 200,000 books, retrieving any segment of it in just over 12 seconds. And, as we saw in the release of the American hostages from Iran, billions of dollars can routinely flow by electronic means from one continent to another in a matter of minutes and with superhuman precision.

But as we reap the benefits of these incredible inventions, others with inventive minds are finding far easier ways to fill their pockets through piracy and other illegal uses of this same technology. The Siemens Corporation of Germany markets in the United States what it calls the world's most advanced ultra-high-security business encryption equipment "because a lot of outside people want inside information." In spite of all precautions, computers lend themselves to espionage, fraud, and theft. Says a Canadian media newsletter: "*Multi-million dollar theft in electronic banking* by a breed of new whiz kid criminals is quite likely in the future."[6] The statement might have been more timely if phrased in the present tense.

A story in the *Los Angeles Times* begins with the question, "How could $21.3 million disappear from the nation's 11th largest commercial bank?"[7] Computer fraud experts and federal bank regulators freely acknowledge that there is virtually no way to prevent such things from happening at just about any major bank in the country. Says one high-ranking expert, "Reasonably intelligent people with a modestly sophisticated scheme can subvert or manipulate even the best bank security system. There is no such thing as a fail-safe procedure."[8]

This incident exemplifies a problem that will become increasingly acute in the information society. Daniel Bell has observed that as computers, information banks, satellites, microprocessors, teletext, and cable are tied together in interactive systems, people become passive observers of processes rather active controllers of events.[9] As we entrust our lives to these sophisticated control mechanisms, we begin to doubt our own capacity to carry out these functions independent of the computers. In such an environment, people who know how to manipulate what the computer tells us have the potential of carrying off fraudulent acts of gigantic proportion simply by exploiting our concerns about "human fallibility."

We are already entering upon an era where communication failures are increasing because of the superabundance of information that cannot be adequately assimilated. When overload occurs, there is a tendency simply to tune out, or to rely even more heavily on what the system tells us we should do. By following that path, Bell warns, such things as the early warning missile system could well end up largely in the hands of thinking computers. Closer to home, could it be that our bank balance was not really as large as we thought it was because the computer printout gave us a different figure? And didn't last month's heating charges seem rather inflated? But how can we judge when the meter feeds its results by fiber optics directly into the automated billing system at utility headquarters? While the computer offers to lift many burdens from our shoulders, it may also have the effect of removing things from our hands. In the end, this labor-saving device may leave us less in command of our lives and more under the control of our "helper." As the information age descends on us, perhaps we are closer to writer Bradshaw's vision of a computer-governed future than we dare acknowledge.

Two other aspects of piracy and fraud are worthy of brief mention. The information society may increase the temptation to use the technology available in any manner that serves our interests. The unauthorized copying of books, articles, broadcast programs, and even computer software is likely to increase dramatically. With the tools readily available, it will be easy to alter their content in a manner that no longer makes them eligible to qualify for protection under the copyright law. With the economic incentive for creating information reduced, fewer and fewer computer programmers, writers, artists, and performers will be willing to produce their wares—ironically, at the very time when channels are proliferating and demand for new material is greater than ever.

During the Joseph McCarthy period, a composite faked photograph was widely circulated. It showed Senator Millard Tidings with Earl Browder of the Communist party. The effort to smear Tidings had only limited success, but it illustrates the potential for the manipulation of images and media content, which will be multiplied tenfold in the approaching electronic age. The possibility of altering or reproducing voice prints (used in computer security systems), of re-recording pictures and sounds with undetectable deletions or additions, and of fabricating authentic-appearing messages that never existed through sophisticated electronic editing techniques raises profound questions about the integrity of future communication and the degree to which they will be trusted and accepted.

III

In an essay on press freedom, *Time* editorialized that "peace and understanding were supposed to follow once the world was wired together into one global village. . . . What the world saw together it would feel together."[10] Yet the global network does not seem to have reduced world tension. The events in Iran serve to illustrate how far we remain from that ideal. As Mosettig and Griggs report, "The Iranians talked to the American government and public via the three privately owned U.S. networks, but they did not make their state owned TV system available for Americans to talk to Iranians."[11]

That "information is power" is an axiom beyond dispute. While America postured about its international commitment to sharing more information with the less developed countries—thus helping to democratize the world—its domestic policies only hardened the argument that information is a commodity to be protected as private property while being bought and sold for a profit. Little wonder that Third World nations doubted our motives and resented the business practices of communication multinationals whose interests we protected.

On the one hand, media spokesmen cry out that freedom of information is indivisible, that no nation has a right to obstruct the process of inquiry or the free flow of news across national boundaries. On the other hand, they demand the right to charge a fee to users of their press and satellite services and to consider information a commodity to be sold to advertisers and cable systems, as well as exported at a handsome profit.

It is clear that some editors and TV management figures are more interested in ratings (or circulation figures) than in the substance on which press freedom is justified. In their search for controversy, broadcast news services are sometimes guilty of creating crises that do not exist. Given the emphasis on instantaneous reporting and beating the competition to the airwaves, there is a measure of superficiality about overseas news that was not always true in the presatellite era. Reporters sometimes lack the language skills, cultural background, analysis time, and depth of understanding needed to give audiences a balanced picture of daily events.

Of even greater concern to those who justify press freedom principles on ethical or moral grounds is the trend to blend news and sensationalism in irresponsible ways. Weekly TV program listings offer evidence of local news programs attempting to compete with titillating movie fare on the latter's home turf. Opposite half-page ads for X-rated movies, one finds lurid pictures announcing coming attractions on the 6 o'clock news. Another example is CBS's public service offering entitled "Older Women, Younger Men." One local station ran a three-quarter-page suggestive photo for its special report on "Herpes Epidemic: The Secret Spreads." One wonders to what extent First Amendment guarantees are appreciated and understood by some news executives in the broadcasting industry.

For all its vaunted freedom, America's media image in the vanguard of investigative reporting and the protection of human rights may be somewhat overblown. While Vietnam was admittedly television's first real war (a war whose end was no doubt hastened by public opinion), it was really the French, Japanese, and Australian TV crews that gave us the early doubts. Without public exposure of that "other" point of view, network news editors might well have been content to follow the Pentagon line without questioning its motives. Except for the self-interest of a gentleman named McCord who spilled the beans at a critical juncture in the *Washington Post's* detective game with government bureaucracy, the dirt of the Watergate affair might easily have been swept under the Oval Office carpet and kept permanently hidden from public view. When we grasped at anything we could get to help raise the curtain of secrecy about the hostages in Iran, the militants orchestrated what they wanted us to hear and we used it in place of real news. Perhaps there is cause to ponder how wisely and how bravely America's press is exercising its freedom.

IV

I have already alluded to the tension between property rights and human rights in the matter of information usage. Many believe that an information society cannot be organized around the concept of private property because information is inherently different from industrial goods. The manufacture of conventional products is dependent on centralization and on mass production involving large investments in plants and equipment. Economies of scale are a function of bigness, of access to raw material, and of distances from the marketplace. The information economy, on the other hand, is almost totally independent of distance or raw materials and is accessible to almost anyone. Both in Canada and the United States, it is estimated that more than half the work force earns its keep by the manipulation of symbols rather than the production of things. As robot technology assumes a larger and larger role in manufacturing, the need for human labor will drop even more drastically. In its place, some foresee a society made of almost infinite cottage industries operating from the home and utterly decentralized in character.

Even as we move inexorably away from the industrial age toward one dependent on the generation of data and the marketing of services, we have yet to resolve the conflict over who shall own or have access to information, let alone how it can best be used for the welfare of the nation. What public policies will govern what may be copyrighted, what rights protected, what access granted, what agencies regulated, and whether they should be licensed at the federal or local level? What principles will guide our decisions about the rights and limitations of satellite consortia, broadcasters, telephone companies, cable operators, newspapers, magazines, IBM or Xerox? What distinctions will prevail if each is purveying data to all or only to some, and if several share the same cable, wire, or optical fiber? Government policy for creating an incentive system with competition is tied to the concept of information as private property. Is that rationale also appropriate for the future, or will the conflict between openness and restrictiveness of information require a more innovative solution?

Nowhere is the ethical aspect of this question more sharply brought into focus than when government itself is challenged to explain why citizens should not have access to information developed with taxpayers' dollars. Long-standing differences between educators and the Central Intelligence Agency, as well as publication of "exposés" by former agents with access to sensitive information, illustrate the complexity of the ethical dimensions of this problem. But even when national security is not involved, there are interesting questions, as yet unresolved by freedom of information laws, about whether a government agency may keep information private while using it to the disadvantage of individual citizens.

That issue has been the basis of an extremely long court fight pitting Susan Long against the Internal Revenue Service and the Justice Department. Mrs. Long, a Princeton University Research Fellow, sought forty-eight reels of IRS computer tape (used by the government to assess probabilities of cheating on tax returns) in connection with her own research on a different topic. After carrying the battle all the way to the U.S. Supreme Court, she was about to claim the disputed tapes when a last-minute U.S. Court of Appeals injunction again blocked the actual transfer. According to an Associated Press story, the Justice Department claimed that "persons knowledgeable in statistical procedures could use the data on the computer tapes to develop formulas so close to IRS formulas that individuals could shift their claimed deductions around in order to reduce their taxes and avoid an audit."[12] This dispute might well help clarify the extent to which information generated within government by public funds is entitled to the same protection as business trade secrets in the private sector.

In the meantime, fierce competition continues within industry for the right to exploit information. The clash over electronic publishing between two giants of the communication field—newspapers and phone companies—is typical. Each seeks to dominate the profitable field of electronic distribution of information now found in newspaper ads or the yellow pages of the telephone directory. Changes in the Communications Act will be lobbied by the American Newspaper Publishers Association to keep Bell Telephone out of that lucrative sphere. Phone company officials will no doubt fight the issue with all the resources at their command. Cable companies vie with one another in life-or-death struggles to sew up major urban markets, and program producers in Hollywood plot their latest strategies for squeezing competition out of the industry and grabbing away the culture market from public television.

V

One may ask with justification, "What has all this to do with morality or with ethics? That is simply the time-tested, grand old American way of doing business!" The answer has to do with the concepts of ethics and morality. Like all complex ideas, there is more than one notion about their meaning and fundamental nature. According to Immanuel Kant, "Morality is not properly the doctrine of how to make ourselves happy, but how we make ourselves worthy of happiness."[13] A very different perspective is heard from Friedrich Nietzsche: "Morality is the best of all devices for leading mankind by the nose."[14] In the more formalistic terms of the discipline of philosophy, normative ethics concerns what we *ought* to do. In tribal societies there are very few moral issues because the members of the group are conditioned from an early age to know what is expected of them. They have little need for an ethical system.

Ethics arises with the advent of social change. It also becomes important when the group is heterogeneous. Both conditions are highly descriptive of the time in which we live. Through the marvels of modern technology we are succeeding in creating the most heterogeneous society that has ever existed. We are also about to experience an era of social change that will make changes in recent years pale by comparison. For both reasons we must begin to turn toward ethical principles in seeking ways to preserve a future society.

One can easily confuse business ethics with the more broadly defined issues of social ethics and morality, which concern us here. Codes of ethics are devices created by special interest groups like the medical profession, the movie industry, or the newspaper editors to protect themselves against pressures for outside regulation of their affairs. The larger meaning of ethical behavior concerns the protection, not of special interests, but of general interests—those affecting most of us in our roles as citizens and consumers.

It is disquieting to observe that neither government nor business has much of a track record in ethical behavior, for both will have great influence on the new information technology. Perhaps it was for that reason that a system of sanctions was developed to offset the excesses of a free market economy. Regulation is, fundamentally, a political process that controls things that do not adequately control themselves. In the late nineteenth century, a number of advocates of the free market society found that the human consequences of the free market were becoming unbearable. In England the plight of the miners, textile workers, and others led those in government to conclude that on humanitarian grounds, there had to be some form of intervention to counter social injustice. The result was the enactment of social legislation.

It is not much in vogue to talk of the "public interest." But one cannot speak of the meaning of morality or ethics without coming to grips with the rights of the individual. Under what circumstances are inequities to be tolerated in a society? In *A Theory of Justice*,[15] philosopher John Rawls speaks of "the liberty principle of ethics," which suggests that each of us has a right to that amount of liberty which is consistent with an equal amount of liberty for everybody else. The second part of this thesis (called the autonomy principle) says that departures from equality are tolerable only when the least-well-off person is better off than he otherwise might have been. Since the goal of the philosopher of ethics is simply to clarify the grounds for what one *ought* to do, the liberty and autonomy principles offer a helpful way of reviewing what the information society may hold in store for us.

VI

In the United States a free market economy operates with the fundamental assumptions of maximum competition among private-sector organizations. There is little reason to believe that this will not be true in the future. Although there are currently 733 commercial TV stations on the air, only three companies provide 90 percent of TV programming in the commercial sector. The Federal Communications Commission, nonetheless, continues to argue that the way to increase competition is to deregulate the industry. Since regulation is the citizen's sole means of exerting influence on the station to act in the public interest, convenience, and necessity, such a policy serves only to weaken the consumer's liberty and autonomy. So in terms of purely ethical principles one would have to question the wisdom of the rush toward decontrol, which reduces protection for the individual and favors corporate property rights.

Deregulation is also justified by FCC Chairman Charles Ferris on the grounds that it encourages diversity within the industry. An examination of the weekly TV program guide reveals how little difference there is in commercial programming from one station to the next. But nothing in the FCC regulations on broadcasting can be blamed for the uniformity of TV content. Nor can we necessarily expect that the future proliferation of channels or the advent of videodiscs will greatly alter the balance of program fare provided. We know from the hard lessons of the past that the media software that emerges never seems to live up to the promise of the new technology for distributing it. As TV critic Arthur Unger puts it, "The main problem has always been, and is going to be, to create worthwhile shows to fill all those long, empty hours."[16] Adds Samuel A. Simon, of the National Citizens Council for Broadcasting, regarding deregulation, "Our job is to empower citizen choice, and we certainly don't think the marketplace gives it."[17]

One can surmise how little liberty or autonomy there will be for the individual citizen in the alleged free market of the information society by the saga of an organization named Premiere Pay Television. In December 1980 the U.S. Federal Court intervened to stop four of Hollywood's largest film studios from forming an alliance with the Getty Oil Company that was designed to keep the three dominant pay-TV competitors from bidding on any of their new film products until nine months after Premiere had offered them to their clients. Stuart Evey, Getty's vice president of diversified operations, publicly complained that operating without an assured advantage in the release of new studio films would be of no interest to him at all. "The cost of building a business without an edge," said Evey, "is way too high a risk to take. . . . The return on investment that way (probably two hundred or three hundred million dollars) wouldn't be worth it. Why invest that much money just to be another store on the block."[18] American "free market" principles indeed! Said FCC Chairman Ferris in a speech to a Los Angeles audience, "A phrase only half jokingly heard at the

FCC is 'All I want is a fair advantage.' There is a natural tendency for all of us to say, 'I'm aboard, now pull up the ladder.'" While that may seem natural by the ethics of business, it hardly lives up to the broader social principles involving opportunity for all to compete on an equal footing for the good life.

Insofar as the new technological age provides special advantage to some in cornering the information market, the possibility exists that, for others, individual liberty and a sense of personal autonomy may be in serious jeopardy. Control of information gives one power to define the issues and develop a consensus in support of them. Said Thomas Jefferson, "If a nation expects to be ignorant and free, ... it expects what never was and never will be."[19] Though he defended press freedom throughout his life, Jefferson warned in 1807, "Perhaps an editor might begin a reformation in some such way as this. Divide his paper into four chapters, heading the first, Truths; second, Probabilities; third, Possibilities; fourth, Lies. The first chapter would be very short."[20] What concerns might Jefferson have had about the potential for manipulating opinion in the age that is now approaching? With two-way interactive cable in every home, instant national plebiscites could be held on any issue, charging congressmen to vote according to the results. One wonders with what wisdom such a public opinion poll might have charted our government's response to the capture of the American Embassy in Iran.

As more and more aspects of our lives are brought within the control of telecommunication systems in the home, there is a danger that we will accept the image on our tube as the actual event—indeed, that we will prefer the symbol to the reality. In hospitals, prisons, and rest homes all across the land, television is already functioning as a pacifier and immobilizer. As the real-world referent becomes blurred in our mind, what is to prevent those who control the source of information from feeding anything they choose into the system? As one who has labored many an hour at the film editing bench, I can attest to how simple it is to create any reality I want from the material at hand. On whose reality will we be basing crucial decisions in the future?

Although their shape may be different and their outline less clear in the future, many of the problems to which I have referred in this chapter have their roots in the present. The coming of the information society will simply compound them and accelerate their impact on our lives. Hegel once wrote that "Hell is truth seen too late." While time remains, there is much to learn about how ethical and moral precepts may help us survive what lies ahead.

Notes

1. From *New York Magazine*, April 19, 1976, pp 63–65.

2. In a speech to Town Hall of California, a Los Angeles civic organization, on January 27, 1981.

3. Bertrand Gross, *Friendly Fascism*, (New York: Evans and Company, 1980), p. 271.

4. Interview by Arthur Unger, TV correspondent for the *Christian Science Monitor*.

5. Joseph Meyers, "Crime Deterrent Transponder System," *Transmission on Aerospace of Privacy*, July 26, 1966.

6. *The Report*, November 14, 1980, p. 5.

7. *Los Angeles Times*, February 7, 1981, part 2, p. 1.

8. Ibid.

9. "The Matching of Scales," Lewis G. Cowan Lecture, International Institute of Communications, London, 1979.

10. Thomas Griffith, "Darkness in the Global Village," *Time*, October 6, 1980, p. 61.

11. Michael Mosettig and Henry Griggs, Jr., "TV at the Front," *Foreign Policy*, no. 38 (Spring 1980): 68.

12. *Los Angeles Times*, February 7, 1981, part 1, p. 10.

13. Kant, *Critique of Practical Reason*.

14. Nietzsche, *The Antichrist*.

15. John Rawls, *A Theory of Justice* (Cambridge: Harvard University Press, 1971).

16. *Christian Science Monitor* series on "Tomorrow's TV," November 4–7, 1971.

17. Ibid., November 7, 1980, p. 13.

18. "U.S. Court Halts Movie Studio's Pay TV Project," *Los Angeles Times*, January 1, 1981, part 1, pp. 1, 22.

19. To Colonel Charles Yancry, 1819.

20. To J. Norvell, 1807.

Chapter Two

Information for What Kind of Society?

Herbert I. Schiller

By and large, communication research in the United States is quixotic. Researchers study intently a world that does not really matter while ignoring the one that does. Much of the work that is done *seems* to be dealing with reality. Communication research concerns itself, among other matters, with development, literacy, broadened participation, and new modes that may provide greater choices. There are also studies of electronic utopias and ecologically inspired colonies in space, utilizing telecommunication to explore personal realms and distant galaxies.

Almost always present is the implicit assumption that things will turn out well. The road is the right one, and the direction taken is promising. Sometimes, in brief periods of doubt, it is admitted that there may be a detour or a few potholes up the road a bit.

"The Information Society: Social and Ethical Issues"—the title of a 1981 symposium—may be symptomatic of the general condition. The major concern of the symposium was expressed in an initial statement by the conference organizer: "A communications revolution, such as is taking place at the present time, will have a tremendous impact on our society and will create certain social and ethical problems."[1]

The statement assumes that if we're not careful, new developments in telecommunications may have unintended and potentially harmful impacts and consequences *in the future*. It is useful, according to the symposium's organizer, to consider these possibilities as early as possible.

In general, such anticipatory thinking is unusual and deserves credit. But in this instance the focus on the future may create more problems than it solves because it does not take account of current reality but shifts attention away from it. The question to be considered is not what *may* happen if a new communication technology runs amok. The issue is much more stark and immediate. It is, How do we check a communication technology that is already running amok and that has had this tendency to do so built into it from the outset? In short, how do we deal with telecommunication systems that have been conceived, designed, built, and installed with the primary objectives being the *maintenance* of economic privilege and advantage and the *prevention* of the kind of social change that would overturn and elimate this privilege?

Developments in telecommunications since the Second World War, with few exceptions, have satisfied these negative objectives. Yet it is amazing how little this has been perceived by those whose task it is to chart and analyze such phenomena. The prime example, and the one most pertinent to this chapter, is the vast military outlay that has been made by the U.S. government over the last forty years. Communication technologies have received a great and growing share of these expenditures.

The attention and favor extended to telecommunications—now institutionalized in an information industry—have not been due to chance. Still it is puzzling that the armaments boom and its runoff on communication technology escape the attention and concern of researchers and workers whose area of interest is at the center of these developments. It is hardly a secret that the armaments output in America has reached staggering levels.[2] Nobel laureate Hannes Alfven suggests that "perhaps it is appropriate to give the *Bulletin* [*of the Atomic Scientists'* doomsday] clock a third hand which would show the seconds we have left in the count down...."[3]

Despite this, each new administration, following the course of previous administrations, has discovered a giant gap in American military preparedness. The ante is being raised once again. In the five-year period from 1981 to 1985, more than $1 trillion dollars ($1000 billion) will be poured into military expenditures—on top of past trillions. A good chunk of this astronomical outlay will go to the production of new telecommunication systems.

Simon Ramo, founder and currently director of TRW, one of the corporate recipients of the population's tax money for military electronics products, explains the impact of the planned military expenditures on the information and communication sector:

> Our military expenditures will probably increase as a percentage of the GNP over the next decade, and this could be extremely pertinent and beneficial to our leadership position in information technology, especially in computers and communications. This is because the right way for us to enhance our military strength in weapons systems is through superior communication, command, control and overall utilization of our weapons systems—functions that depend on superior information technology.[4]

Unlike most communication researchers, Simon Ramo has the priorities in their proper order.

Actually, what Ramo is anticipating has been the practice throughout most of the past four decades. The entire electronics industry is an outcome of military subsidy and encouragement. The early computers and all their successors were developed and built with government funds. Illustrative, "the ILLIAC, the largest and most advanced computer of the day [1960s], was developed with funds from the Advanced Research Projects Agency of the Department of Defense."[5]

The first communication satellite system was a military effort. Most of the satellites in the sky today are providing military data. In January 1981 *Science* reported that "the Army soon expects to have 13,000 computers, the Navy, 33,000, and the Air Force, 40,000. The software bill for military computers last year came to more than $3 billion."[6]

So great is the Department of Defense's utilization of this huge arsenal of computers that it is compelled to finance the development of a standardized computer language—Ada—to overcome its present reliance on more than 1000 computer languages. But in this area as well, military influence has had a long history. "The Pentagon," *Science* reminds us, "was the driving force behind the development of COBOL (Common Business-Oriented Language), which was introduced in 1959 and is today use extensively around the world."[7]

Still, the development of computer languages is but one debt the information age owes to its military parentage. Much more direct, though far less accessible than Ada or COBOL, is the dense U.S. electronic espionage network that encircles the world, not excluding national territory. This, too, is a creation made possible by the most sophisticated communication technologies.

The supersecret National Security Agency (NSA), for example, works largely with satellites, microwave stations, and computers. "Its mission includes cracking enemy codes, developing unbreakable ciphers for the United States and, most importantly, monitoring, translating and analyzing worldwide communications among nations, selected foreign citizens and some corporations."[8]

Australia is practically an NSA preserve. The country serves as a continental base, across the Pacific Ocean, for a full complement of NSA electronic installations, to be directed against the Soviet Union and any other perceived adversary. It is described by one Australian reporter as a "massive eavesdropping on Australian communications . . . conducted by facilities dotted around Australia as part of a U.S. nuclear war fighting machine that has been nurtured on Australian soil by [an] inner circle."[9] Desmond Ball, who documented the U.S. electronic presence in Australia, inscribed his book "For a Sovereign Australia."[10]

Australians are not alone in having their communication—personal, private, and governmental—intercepted and monitored. The NSA performs the same functions with abandon around the world including inside the United States. A NSA intercept station outside London, for example, checks all incoming and outgoing U.K.-European communication traffic. The annual budget for the operation of this global assemblage of modern communication technology is a secret. According to the *New York Times*, "Intelligence officials estimated the agency's budget to be more than $2 billion a year, larger than that of the Central Intelligence Agency."[11] What does this huge sum buy? Nothing less than what many generously call the "information society."

According to a report of Charles Morgan, who used data supplied elsewhere by Harrison Salisbury, in 1973 alone, "NSA retrieved a total of 23,346,587 individual communications. . . " Morgan notes with astonishment that this figure is "not the number of individual communications *intercepted*; it is the number of communications *retrieved* for full study by the NSA."[12]

By 1980 the existence of the National Reconnaissance Office had been revealed. Still another supersecret military agency, its mission "is to oversee the development and operation of spy satellites used to photograph foreign territory and to monitor international communications."[13] Its budget exceeds $2 billion a year. Its satellites in the 1960s and '70s were used "to photograph antiwar demonstrations and urban riots. . . ."[14]

The CIA deserves mention at this point. It is no wallflower in the business of electronic surveillance. Though it is no longer fashionable to bring up Watergate, it will be remembered that the crew of bunglers recruited by the Nixon high command came out of the agency.

In sum, a great amount of the activity, a good share of the content, and the general thrust of what is now defined as the information age represent military and intelligence connections.

Scholarly work in telecommunications puts these matters differently. In what has become a familiar argument, the duality and ambiguity of the new communication instrumentation is emphasized. The potential for constructive as well as destructive ends is insisted upon. For instance, Anthony Oettinger, director of Harvard's Program on Information Resources Policy, writes:

> Tensions will continue because of the close kinship among the technical means used for sometimes antithetical purposes—keeping peace or waging war, gathering intelligence or mapping agricultural and mineral resources, providing communications for trade or for military command and control. . . .[15]

But is the new instrumentation truly indifferent to the uses to which it is applied? Can it be utilized for social and peaceful goals as well as its current applications to war and war preparations?

Given the prevailing structure of global industrial and military power, it is difficult to believe that the communication revolution is not the outcome of deliberate and extensive efforts to maintain a worldwide system of economic advantage. New information technologies have been invented, developed, and introduced to support the business component of this system and to enable a globe-girdling military communication network to be prepared to be the ultimate enforcer. These are the objectives of those who were instrumental in the development of the telecommunications industry. These objectives account in large measure for its genesis.

As Deputy Assistant Secretary for Human Rights and Social Affairs Sarah Goddard Power explained these dynamics:

> Either we will design, produce, market and distribute the most advanced products and services spun off by the communications revolution—and, in so doing, reinforce our economic as well as political, social, and cultural *advantage*—or we will increasingly find ourselves in the position of consumer and debtor to those who do. . . . The question of how the world adapts to the communications revolution has been steadily moving up the list of international concerns over the last decade, and it has now emerged as a major point of contention in East-West and North-South relationships.[16]

Maintaining and "reinforcing our advantage" are the explicit grounds that activate and accelerate the communication revolution. There is no ambiguity or dualism here. It is not a question of "either-or"—good technology use or bad technology use. It is solely a matter of developing and using the new communication technology for holding on to the economic benefits derived from a world system of power. For this reason, insistence on the potential and positive features of the current communication instrumentation is disingenuous at best.

To be sure, a part of the technology is used to produce consumer products, games, and "entertainment." Many of these are actually applications seeking a use. Assorted privateers of enterprise naturally make the most of the opportunities created by this technology.

It is also undeniable that the commercial utilization of telecommunications will affect employment in the private and public sectors. The character of work itself will change. Living patterns will alter. House, home, and family arrangements will change. What will not be different—not if the corporate-military directorate has anything to say about it—are basic relationships of authority, ownership, and hierarchy of skills.

All the same, domestically and internationally, the evolving telecommunications industry is being hailed and promoted as efficient, problem solving, and liberating. The capability for enormously expanded generation and transmission of volumes of information, the technical feasibility of two-way communication, and the choice and diversity that new information technologies allegedly will provide, are publicized widely as realizable goals, affording hope to disadvantaged and excluded nations, classes, and people. These are not only unrealizable objectives. Under prevailing circumstances, their enunciation is misleading and deceptive. Other than for a few meritorious functions (libraries, medicine).[17] significant socially beneficial utilization of the new technologies requires societal restructuring. The notion that humanistic social change can be introduced incrementally, via the new technology, is unrealistic to the point of fantasy.

The social potential that may exist, and I stress the "may," in some of the new instrumentation can be developed appreciably only in totally different social-cultural-economic contexts. Claims, therefore, that the two-sidedness of the instrumentation—its potential for good or evil—necessitate its immediate development in the hope that the socially desirable side may be encouraged are either uninformed by history or, more likely, too well informed by special interest.

In the meantime, the world, especially the U.S.-administered part of the world, is being hooked into electronic circuitry that serves to keep things under control for the transnational corporate community of IBM-Chase-Manhattan-Citicorp-Exxon-Arco-CBS-J. Walter Thompson and their friends. Transnational data flows within and among these business giants have become indispensable to the maintenance of the world business system. Additional networks thicken the connections and extend the system's influence.[18] The stronger the electronic networks established, the less likely the possibility for national autonomy and independent decision making, and the more likely an intensified patroling and controling of the Third World majority.

Surveillance, monitoring, and marketing are the near-certain outcomes of the utilization of new communication technologies, domestically and globally. The American public, half of which does not bother to vote in national elections, hears the good news that electronic referendums are around the corner. Sitting at home, in front of a domestic information utility, so-called, the happy citizen will be able to exercise innumerable inconsequential choices on an electronic console in the living room. This, we are told, constitutes the most advanced form of democracy.

Actually, business and marketing, law and order, and war are the main progenitors of telecommunications and the chief users of the advanced systems and processes. Yet all these developments could scarcely have reached their current psychotic levels without the assistance of still another vital component in the national communications system: the mass media. The media use many forms of telecommunications in the production of their services—news, entertainment, drama, music, film—and this is crucial, they interpret for all of us why we are supposed to need the new technologies. In a conflict-of-interest situation that "boggles the mind," the print and electronic media, more often than not, are part of larger conglomerates in the new information industries. Yet in their capacity as information providers they instruct their readers/viewers/listeners on why massive armaments are good for everyone, why the demands of the poor (people and nations) are not to be taken seriously, and why the United States must "lead" the world.

In providing these guidelines, they treat all of us daily to belligerent bulletins, some of which could be primary documentation—if we survive—for a future war crimes tribunal, hearing testimony about our media gatekeepers. An issue of the *Columbia Journalism Review*, for example, carried an excessively cautious article, which nonetheless detailed the fondness of the press for presenting nuclear war to its readers and listeners and viewers as a feasible, controllable option.[19] In this particular account, the responsibility for nuclear war mongering is shifted to the shoulders of reporters, who are described as enamored with the jargon of Pentagon briefers. This may have helped to get the article printed, but it will not do as an explanation of why lunatic reportage of this character is published. War, even nuclear war, has become an option, to be used to hold on to a transnational empire, many components of which are becoming unstuck.

All of this means that "the information age" is a misnomer. So too is "the telecommunications revolution." A few advanced industrial societies are striving to assure their privileges in a revolutionary world in which at least 3 billion people no longer are accepting quietly their long-standing conditions of exploitation and servility. Information systems have been developed to maintain—albeit in new ways—[20] relationships that secure the advantages enjoyed by a small part of humanity and the disadvantages that afflict the large majority.

Consequently, it is a mistake to believe that the changes required to overcome the global, national, and local disparities in human existence will be facilitated by developing telecommunication systems. In fact, the opposite result may be expected. Existing differentials and inequities will be deepened and extended with the new instrumentation and processes, despite their loudly proclaimed and widely publicized potential benefits. Only *after* sweeping change inside dozens of nations, in which ages-old social relationships are uprooted and overturned, can the possibility of using new communication technologies for human advantage begin to be considered. It can be taken for granted, also, that the technologies applied at such a time hardly will approximate those now in use and in operation.

How would communication researchers and information workers in general relate to this situation? Actually, their contribution could be a worthy one. It would require serious social and ethical consideration of the new communications technology. It would require, however, a totally different stance from what has come to be expected of communication research and allied information worker community. Historically, communication as a field of endeavor has served the interests of the industrial marketing community, the psychological branch of the military, and the information arm of the governing bureaucracy, especially its international side. In effect, these are core elements in the U.S. domestic and global empire. Unquestioning allegiance to these power centers is tantamount to a denial of social and ethical responsibility.

Accordingly, to be socially responsible, communication workers, theoretical and practical workers alike, must distance themselves, as best they can, from the centers of power. This is no simple or easy task. It means, first and foremost, to challenge the central sources of communication control and distortion in the country. It requires, for example, that the unprecedented arms buildup in the country be understood and then explained by communication people, wherever they have an opportunity to do so, as an effort to maintain the global status quo, as a way of denying resources and political autonomy to much of the world, and as the main source of the new electronic technologies, which are applied to governance, coercion, and control.

The social forces and institutional structures that produce the need for nuclear weapons are the proper subjects for research, analysis, and presentation. The connection of nuclear arsenals and plants of physical annihilation to the supportive informational system—the mass media—requires no less attention and debate.

In sum, what telecommunications as a field has as its domain of research, study, and subject matter for popular presentation, if it would accept its social responsibility, is nothing less than the physical, structural, and institutional bases of a system of domination that operates through its impact on human consciousness. No one could say that this is an easy assignment. The interlocking networks of power and authority that have grown up in this country in the twentieth century make an adversary position disagreeable and difficult, and perhaps dangerous. Even the small space for an independent critique that once existed in the academic realm—and it was a small space, to be sure—is being recaptured by a rapidly forming corporate-university entente.

But the subject at hand is social and ethical responsibility. Is there any viable choice other than one of critical opposition to the foundations of what is now hailed as the "information society"?

Notes

1. J. L. Salvaggio, "Statement of Purpose," Edward R. Murrow Symposium, July 7, 1980.

2. Such widely different observers as E. P. Thompson, the distinguished English historian, and the Reverend Billy Graham share anxiety over the level of military spending and the amount of stockpiled weapons.

3. Hannes Alfven, "Human IQ vs. Nuclear IQ," *Bulletin of the Atomic Scientists*, January 1981, p. 5.

4. Simon Ramo, *America's Technology Slip* (New York: Wiley, 1980), p. 283.

5. Gina Bari Kolata, "Who Will Build the Next Supercomputer?" *Science* 211 (January 16, 1981): 268–69.

6. William J. Broad, "Pentagon Orders End to Computer Babel," *Science* 211 (January 2, 1981): 31–33.

7. Ibid.

8. Philip Taubman, "Choice for C.I.A. Deputy Is an Electronic-Age Spy," *New York Times*, February 2, 1981.

9. Brian Toohey, "How Australians Are Kept in the Dark (While the U.S. Listens In)," *National Times* (Australia), November 16–22, 1980.

10. Desmond Ball, *A Suitable Piece of Real Estate* (Sydney, Australia: Hale & Iremonger, 1980).

11. Taubman, "Choice for C.I.A. Deputy."

12. Charles Morgan, "The Spies That Hear All," *San Diego Newsline* 4, no. 11 (December 17–24, 1980). (Emphasis in text.)

13. Philip Taubman, "Secrecy of U.S. Reconnaissance Office Is Challenged," *New York Times*, March 1, 1981.

14. Ibid.

15. Anthony G. Oettinger, "Information Resources: Knowledge and Power in the 21st Century," *Science* 209 (July 4, 1980): 191–98.

16. Sarah Goddard Power, *The Communications Revolution*, Current Policy No. 254, United States Department of State, Bureau of Public Affairs (Washington, D.C.: Government Printing Office, December 5, 1980). Emphasis added.

17. And these require close scrutiny as well.

18. Herbert I. Schiller, *Who Knows: Information in the Age of the Fortune 500* (Norwood. N.J.: Ablex, 1981).

19. Fred Kaplan, "Going Native Without a Field Map," *Columbia Journalism Review*, January/February 1981, pp. 23–29.

20. Schiller, *Who Knows*.

Chapter Three

Communication Technology— for Better or for Worse?

Daniel Bell

Human societies have seen four distinct revolutions in the character of social interchange: in speech, in writing, in printing and, now, in telecommunication. Each revolution is associated with a distinctive, technologically based, way of life.

Speech was central to the hunting and gathering bands—the signals that allowed men to act together in common pursuits. Writing was the foundation of the first urban settlements in agricultural society—the basis of recordkeeping and the codified transmission of knowledge and skills. Printing was the thread of industrial society—the basis of widespread literacy and the foundation of mass education. Telecommunications (from the Greek *tele*, "over a distance")—the ties of cable, radio, telegraph, telephone, television, and newer technologies—is the basis of an "information society."

Human societies exist because they can purposefully coordinate the activities of their members. (What is a corporation if not a social invention for the coordination of men, material, and markets for the mass production of goods?) Human societies prosper when, through peaceful transactions, goods and services can be exchanged in accordance with the needs of individuals.

Central to all this is information. Information comprises everything from news of events to price signals in a market. The success of an enterprise depends in part on the rapid transmission of accurate information.

The foundation of the Rothschild fortune was advance information by carrier pigeon of the defeat of Napoleon at Waterloo, so that the Rothschilds could make quicker stock market decisions. (The rapidity of transmission of information on companies today is responsible for the random walk theory of stock market prices, since such rapidity minimizes the time advantage of inside information.)

General equilibrium theory in economics is dependent on "perfect information," so that buyers and sellers know the full range of available prices on different goods and services, and the markets are cleared on the basis of relative prices and ordinal utilities. What was once possible by walking around a local market now has to be done through complex transmission of news, which flashes such information to clients in "real time."

All facets of society are concerned with information—from balance-of-trade figures to money supply, birthrates, interregional shifts, changes in buying tastes and habits—should recognize more readily than any other the importance of any changes in the type and character of information. For this reason, we should understand the nature and extent of the powerful technological revolution that will escort us into the information era, as well as its potentialities and threats.

A New Communication System

In the 1980s, more than 130 years after the creation of the first effective telecommunication device, telegraphy, we are on the threshold of a new development that, by consolidating all such devices and linking them to computers, earns the name of a "revolution" because of the various possibilities of communication that are now unfolding. This is what Simon Nora and Alain Minc, in an extraordinary report to the president of the French Republic, call *telematique*, or what Anthony Oettinger calls *compunications*.[1]

Telematique, or *compunications*, is the merging of telephone, computers, and television into a single yet differentiated system that allows for transmission of data and interaction between persons or between computers through cables, macrowave relays, or satellites. Thus communication becomes faster, but it also is organized in a totally new way. It would be far beyond the scope of this essay to specify these ways in detail, but it is possible to suggest some of the basic communication modes in an information society and illustrate the consequences.

Data processing networks. These would register purchases made in stores automatically through computer terminals as bank transfers. Orders for goods, such as automobiles, would be sent through computer networks and transformed into a programming and scheduling series to provide for individual specifications of the items ordered. In a broad sense, this could be a replacement of much of the "paper economy" by an electronic transfer system.

Information banks and retrieval systems. In an information society these would recall or search for information through computer systems and would print out a legal citation, a chemical abstract, census data, market research material, and the like.

Teletext systems. In these systems, such as the British Post Office Prestel system (formerly called View Data), or the French Tic-tac and Antiope systems, news, weather, financial information, classified advertisements, catalog displays, and research material are displayed on home television consoles, representing a combination of the yellow pages of telephone books, the classified advertisements of newspapers, standard reference material, and news.

Facsimile systems. Here, documents and other material (invoices, orders, mail) can be sent electronically rather than by postal systems.

Interactive on-line computer networks. These allow research teams or office managers or government agencies to maintain communication so as to translate new research results, orders or, perhaps, financial information into further action.[2]

These are not speculations or science fiction fantasies; they are developed technologies and will be central to information societies. The rate of introduction and diffusion will vary, of course, on the basis of cost and competition of rival modes, and on government policies that will either facilitate or inhibit some of these developments.

The rate of diffusion is further compounded by capital problems: The need for a large-scale shift to new, independent sources of energy requires a large, disproportionate allocation of capital to purposes that are, inherently, "capital-using" rather than "capital-saving." Thus the marginal efficiency of capital (as reflected in social capital-output rations) tends to fall. The uncertainty of inflation leads, sometimes, to the postponement of capital investment or the short-term substitution of labor inputs rather than capital inputs, thus dragging down further a society's total productivity. These are economic and political questions that, again, are outside the scope of this article.

If we assume, however, that many of these new technologies and modes will eventually be introduced, what can we say of their consequences? It is hazardous, if not impossible, to predict specific social changes and outcomes. What one can do is to sketch broad social changes that are likely to occur when these new modes are all in use. And that is the purpose of the following two sections.

Societal Infrastructure

Every society is tied together by three different kinds of infrastructure—transportation, energy grids, and communication:

Modes of Transportation

The oldest of these infrastructures is transportation, which first took place by trails, roads, and rivers, and later by canals. Trade was the means of breaking down the isolation of villages and served as a means of communication between distant areas. Transport thus has been the major linkage between settled areas.

Because of transport requirements, all the major cities of the world have been built near water. The industrial heartland of the United States, for example, was created by the interplay of resources and water transport.

Thus the iron ore from the Mesabi Range could move on Lake Superior, and coal in southern Illinois and western Pennsylvania could be tied to the Great Lakes by a river system. Such a network allowed the development of a steel and then an automobile industry. The water transport system served to thread together the industrial cities of Chicago, Detroit, Cleveland, Buffalo, and Pittsburgh.

In Germany, in the early eighteenth and nineteenth centuries, most commerce flowed from north and south because of the course of the major rivers such as the Rhine, the Elbe, the Oder, and the Weser. The coming of the railroad, linking east and west, greatly facilitated the unification of Germany by 1870 and its development as an industrial and military power.

Power Sources

The second infrastructure has been energy. At first waterwheels on rivers were used for power, followed by hydroelectricity, then oil, gas, and electricity.

The interaction of the energy and transport systems allowed for the spread of industries and towns, since electricity grids could transmit power over hundreds of miles. The result was the development of large industrial complexes, occupying vast spaces, through the long-distance transmission of energy.

Communication Systems

The oldest communication infrastructure is the postal service. Much later came the development of the various telecommunications systems.

In an information society it is likely that there will be a major shift in the relative importance of the infrastructures: telecommunications will be the central infrastructure tying together a society. Such a network increases personal interaction and drastically reduces the costs of distance. It affects the location of cities, since the "external economies"—gains because of proximity, such as in advertising, printing, and legal services for banks—once possible only in central city districts are being replaced by communication devices.

Most important, an information society enlarges the arenas in which social action takes place. It is only since 1950 that many countries, because of revolutions in air transport and communication, have become national societies, in which impacts in any one part of the national society are immediately felt in any other part.

In the broadest sense, we have for the first time a genuine international economy in which prices and money values are known in real time in every part of the globe. Thus, for example, treasurers of banks or controllers of corporations can subscribe to a Reuters international money market service and obtain, in real time, quotations on different currencies in twenty-five different money markets from Frankfurt to London to New York to Tokyo to Singapore to Hong Kong, so that they can take advantage of the different rates and move their holdings about.

By satellite communication, through television, every part of the world is immediately visible to every other part. The multiplication of interactions and the widening of the social arenas are the major consequences of a shift in the modalities of the infrastructure. This is a problem we shall return to later.

Postindustrial Society

The creation of an information society also speeds the development of what I have called a postindustrial society.[3] Table 3.1 schematically compares preindustrial, industrial, and postindustrial types of developments.

Most of the world—that is, principally the countries in Asia, Africa, and Latin America—is preindustrial in that 60 percent or more of its labor force is engaged in extractive industries. The life of these countries is a "game against nature," in which national wealth depends on the quality of the natural resources and vicissitudes of world commodity prices.

A smaller section of the world, the countries around the North Atlantic littoral plus the Soviet Union and Japan, is made up of industrial countries where the fabrication of goods, by the application of machine technology with energy, is the basis of wealth and economic growth.

Some of these latter countries are moving into the postindustrial world. In the postindustrial state, first there is a shift from the production of goods to the selling of services. Services exist in all societies, but in preindustrial societies they are primarily domestic services. In industrial societies, they are ancillary to the production of goods, such as transportation, utilities, and financial services. In postindustrial societies, the emphasis is on human services (education, health, social services) and on professional services (computing, systems analysis, and scientific research and development).

The second dimension of postindustrial society is more important: the fact that, for the first time, innovation and change derive from the codification of theoretical knowledge. Every society has its base, to some extent, in knowledge. But technical change has now become dependent on the codification of theoretical knowledge. We can see this easily by examining the relation of technology to science.

Steel, automotive, utilities, and aviation industries are primarily of the nineteenth century in that they were created largely by inventors— "talented tinkerers"—who knew little about the basic laws or findings of science. This was true of such a genius as Edison, who invented, among other things, the electric lamp, the gramophone or record player, and the motion picture. Yet he knew little of the work of Maxwell or Faraday on electromagnetism, the union of whose two fields was the basis of almost all subsequent work in modern physics. This was equally true of Siemens with his invention of the dynamo, and Bell with the telephone, or Marconi with the radio wireless.

Table 3.1
The postindustrial society: a comparative scheme

	Preindustrial	Industrial	Postindustrial
Modes			
Mode of production	Extractive	Fabrication	Processing & recycling services
Economic sector	Primary Agriculture Mining Fishing Timber Oil & gas	Secondary Goods producing Durables Nondurables Heavy construction	Tertiary Transportation Utilities Quaternary Trade Finance Insurance Real estate Quinary Health Research Recreation Education Government
Transforming resource	Natural power – wind, water, draft animal-human muscle	Created energy – electricity, oil, gas, coal, nuclear power	Information* – computer & data transmission systems
Strategic resource	Raw materials	Financial capital	Knowledge†
Technology	Craft	Machine technology	Intellectual technology
Skill base	Artisan, farmer, manual worker	Engineer, semiskilled worker	Scientist, technical & professional occupations
Methodology	Common sense, trial & error, experience	Empiricism, experimentation	Abstract theory: models, simulations, decision theory, systems analysis
Time perspective	Orientation to the past	Ad hoc adaptiveness, experimentation	Future orientation: forecasting & planning
Design	Game against nature	Game against fabricated nature	Game between persons
Axial principle	Traditionalism	Economic growth	Codification of theoretical knowledge

* Broadly, data processing: The storing, retrieval, and processing of data become the essential resource for all economic and social exchanges

† An organized set of statements of facts or ideas, presenting a reasoned judgment or experimental result, that is transmitted to others through some communication medium in some systematic form

The first "modern" industry is chemistry, in that the scientist must have a knowledge of the theoretical properties of the macromolecules that he is manipulating in order to know where he is going. What is true of all the science-based industries of the last half of the twentieth century, and the products that come from them, is that they derive from work in theoretical science, and it is theory that focuses the direction of future research and the development of products.

The crucial point about a postindustrial society is that knowledge and information become the strategic and transforming resources of the society, just as capital and labor have been the strategic and transforming resources of industrial society. Th crucial "variable" for any society, therefore, is the strength of its basic research and science and technological resources—in its universities, in its research laboratories, and in its capacity for scientific and technological development.

In these respects, the new information technology becomes the basis of a new intellectual technology in which theoretical knowledge and its new techniques (such as systems analysis, linear programming, and probability theory), hitched to the computer, become decisive for industrial and military innovation.

Corollary Problems

Two important consequences of the revolution in telecommunications round out the picture of social change. One is that, because of a combination of market and political forces, a new international division of labor is taking place in the world economy; the other involves a widening scale of political effects across the world.

Economic Changes

The developing countries, in proclaiming a new international economic order at Lima in 1975, have demanded that 25 percent of the world's manufacturing capacity be in the hands of the Third World by the year 2000. This is a highly unrealistic target. Yet some tidal changes are already taking place.

There is one group of developing countries—among them Brazil, Mexico, South Korea, Taiwan, Singapore, Algeria, Nigeria—that is beginning to industrialize rapidly. It is likely that in the future traditional, routinized manufacturing, such as the textile, shipbuilding, steel, shoe, and small consumer appliances industries, will be "drawn out" of the advanced industrial countries and become centered in this new tier.

The response of the advanced industrial countries will be either protectionism and the disruption of the world economy or the development of a "comparative advantage" in, essentially, the electronic and advanced technological and science-based industries that are the feature of a postindustrial society. How this development takes place will be a major issue of economic and social policy for the nations of the world.

Expanded Political Arena

The second, more subtle, yet perhaps more important, problem is that the revolution in telecommunications necessarily means a change in scale—an expansion in the political arenas of the world, the drawing in of new claimants, and the multiplication of actors or constituencies.

We have heard much of the acceleration of the pace of change. It is a seductive yet, in the end, a meaningless idea other than as a metaphor. For one has to ask, "Change of what?" and, "How does one measure the pace?" There is no metric that applies in general, and the word *change* is ambiguous.

As Mervyn Jones, the English author, once pointed out, a man who was born in 1800 and who died in 1860 would have seen the coming of the railway, the steamship, the telegraph, gas lighting, factory-made objects, and the expansion of the large urban centers. A man who was born in 1860 and who died in 1920 would have seen the telephone, electric lights, automobile, and motion pictures. He might be familiar with the ideas of Darwin, Marx, and Freud. He would have seen the final destruction of most monarchies, the expansion of the ideas of equality, and the rise and breakup of imperialism.

How does one measure the events of the past forty years in order to say that the *pace* of change has increased? If anything, one might say that, since growth is never exponential in a linear way but follows an S-shaped or logistic curve, we are close to "levelling off" many of the so-called changes that have transformed our lives (e.g., transport and communication will not increase appreciably in speed). And in the world increase in population, we seem to have now passed the "point of inflection," that midway point where the S-curve of change is now slowing down.

But what is definite is that the scale on which changes have taken place has widened. And a change in scale, as physicists and organization theorists have long known, requires essentially a change in form. The growth of an enterprise, for example, requires specialization and differentiation and very different kinds of control and management systems when the scales move from, say, $10 million to $100 million to $1 billion.

The problem becomes politically acute for political systems. Rousseau, in *The Social Contract*, set forth a "natural law" that the larger a state becomes, the more its government will be concentrated, so that the number of rulers decreases as the population increases. Rousseau was seeking to show that a regime necessarily changes its form as the population increases, as the interactions between people multiply, and as interests become more complicated and diverse.

The problem for future information societies—especially those that wish to maintain democratic institutions, the control of government by the consent of the people, and an expanding degree of participation—is to match the scales between political and economic institutions and activities. The fact that government has increasingly become more distant from and yet more powerful in the lives of persons has led increasingly, as well, to separatism, localism, and breakaway movements in society.

At the same time, the scale of economic activities on a worldwide canvas has indicated that we lack the governing mechanisms to deal with, for example, monetary problems, commodity prices, and industrial relocation on the new scales on which these actions take place. As I remarked in an earlier article, what is happening is that, for many countries, the national state is becoming too big for the small problems of life and too small for the big problems of life.[4]

Implications for Personal Liberty

All of these structural changes lead to the pointed question of the fate of individual and personal liberties in this "brave new world." We have had, from Aldous Huxley to George Orwell, dire predictions of the kinds of controls—the expansion of Big Brother totalitarianism—that may be coming as a result of such technological changes. Indeed, an old Russian joke asks: Who is Stalin? Answer: Genghis Khan with a telephone. And there are many humorous examples of how the new technologies permit the growth of scrutiny mechanisms and intrusions in personal life.

A story in the London *Times* on the growth of security procedures in Germany reports that the movements of a German business consultant who has to cross the border into Switzerland several times a day were reported to a computer center so that he suddenly found himself on a list of persons to be watched. But the moral of the story is not that the computerization of the border crossings increases the power of the police but that because of the political threats of terrorism, such procedures had to be adopted.

The issue of social control can be put under three headings:

1. Expansion of the techniques of surveillance
2. Concentration of the technology of recordkeeping
3. Control of access to strategic information by monopoly or government imposition of secrecy.

In all three areas there has been an enormous growth of threatening powers and, in a free society such as ours, a growing apprehension about their misuse.

The techniques of surveillance, since they are the most dramatic, have received the most attention. In George Orwell's *1984*, the government of Oceana monitors party members by a remote sensor of human heartbeats. The sensors are located in the two-way television screens in all homes, government offices, and public squares. By tuning in on individuals and measuring their heartbeats, Big Brother can discover whether an individual is engaged in unusual activities.

A young physiologist has already discovered, to his dismay, that he had invented such a device himself. Seeking to measure the physiological activities of salamanders less painfully than by plugging painful electrodes into the animals' bodies, he created a delicate voltage sensor that measures the extremely minute electrical field that surrounds the bodies of all living organisms so that one can now detect, and record from a distance, an animal's heartbeat, respiration, muscle tension, and body movements. Told that he had invented the device that Orwell had imagined, he made a study of the predictions Orwell made in *1984* and found that, of the 137 devices Orwell had described, some 100 are now practical.[5] Also, in Solzhenitsyn's *First Circle*, it may be recalled, the prisoner-scientists in the secret police laboratory were working on devices to identify telephone callers by voiceprints and also to unscramble coded telephone conversations—devices that are in use today.

The computerization of records, which intelligence agencies, police, and credit agencies use, is by now quite far advanced, and so much so that individuals are constantly being warned to check to see that their credit ratings are accurately recorded in computer memories, lest they be cut off, especially in cashless and checkless transactions, from the purchase of goods that they need.

The problem of secrecy is old and persistent. *Science* magazine has reported that at the request of the National Security Agency (NSA), the Department of Commerce imposed a secrecy order that inhibited commercial development of a communication device, invented by a group in Seattle, for which a patent had been applied. The technique involved in the patent application goes beyond the voice-scrambler technology used in police and military communication and takes advantage of the spread-spectrum communication band to expand the range of citizen's band and maritime radios. *Science* reports:

> The inventors are fighting to have the order overturned so that they can market their device commercially. They regard their struggle as a test of whether the government will allow the burgeoning of cheap, secure communications technology to continue in the private sector or whether it will keep a veil of secrecy over the work—effectively reserving it exclusively for military and intelligence applications.[6]

Real as these issues are for liberty in the personal and economic sense, they are not the true locus of the problem. It is not in the technology per se but in the social and political system in which that technology is embedded.

The most comprehensive system of surveillance was invented by that malign individual Joseph Fouche, who served as police chief for Napoleon I. A former Catholic priest, Fouche became a militant leader of the French Revolution, having directed the massacre at Lyons; and after Napoleon he continued as police chief during the Bourbon restoration of Louis XVIII. Fouche was the first to organize every *concierge* in Paris as an agent of the police and to report to local headquarters the movements each day of every resident of the buildings.

The scale of operations has expanded since Fouche's day. Technology is an instrument for keeping abreast of the management of scale. The point can be made more abstractly, yet simply. Technology does not determine social structure; it simply widens all kinds of possibilities. Technology is embedded in a social support system, and each social structure has a choice as to how it will be used. Both the Soviet Union and the United States are industrial societies, using much the same kind of technology in their production systems. Yet the organization of industry, and the rights of individuals, vary greatly in the two societies.

One can take the same technology and show how different social support systems use them in very different ways. For the automobile, one can show very different patterns of use, and consequent social costs, without changing a single aspect of the automobile itself. Thus,

in one kind of society, one can have a system of complete private ownership of the automobile where the individual can come and go largely
at his own pleasure. But such a pattern involves a high cost to the
individual for the purchase of the car, the insurance, gasoline, and de-
preciation, as well as a cost to the community for more roads, parking
garages, and the like.

Yet one can envisage a very different pattern, which some cities
have tried, where automobiles are barred from a large area and in their
place there is a "public utility" system in which an individual sub-
scribes to a car service. He goes a short distance—no farther, say, than
a usual bus stop—to a parking lot, takes a car, using a magnetic-coded
key; drives off in that car to his destination; and simply leaves the car
in that other lot, again no farther than a bus stop distance from where
he wants to go.

The user has a great degree of mobility (as with a taxi service, yet
without the cost of a driver), but fewer cars are needed in this kind of
distributive system. The cost to the individual is the walk to and from
the parking lot and the brief wait for a car that may be necessary if
there are shortages.

Such a system represents an expansion of the car-hire service that is
available at airports and throughout cities in many countries. The illus-
tration is trivial (and I am not arguing for one or another of the patterns to be imposed on a society); yet the import of the example is not
so trivial: A single technology is compatible with a wide variety of
social patterns and the decisions about the use of the technology are,
primarily, a function of the social pattern a society chooses.

To expand this proposition, the following theorem holds: The revolution in telecommunications makes possible both an intense degree
of centralization of power, if the society decides to use it in that way,
and large decentralization because of the multiplicity, diversity, and
cheapness of the modes of communication.

It is quite clear that an elaborate telecommunications system allows
for intensification of what in military parlance is called command and
control systems. Through such systems, the U.S. Air Force was able to
set up a watch pattern that kept track of all aircraft or unidentified
flying objects over the North American air space and relayed that information, in real time, to a centralized control station that in turn
monitored the information.

Without such a system, one could not have basic security against an
enemy attack. Yet, in the Vietnam war, the development of the command and control system meant that tactical decisions, which in the

past were made by field commanders, were often made on the basis of political decisions in Washington. The Vietnam war was an extraordinary instance of the centralization of military decisions on a scale rarely seen before.

In Chile, from 1971 to 1973, the British organization theorist Stafford Beer set up an effort to prepare a single computer program to model —and eventually control—every level of the Chilean economy. Under his direction, with the cooperation of the Allende government, an operations room was created to plan for centralized control of Chilean industry. This was not a simulations model, as used by Jay Forrester and his associates to demonstrate what might be if certain assumptions held, but an operational recursive model (i.e., a set of "nestings" or mini-systems built into a pattern of larger systems) to direct the course of the economy from a single center. Whether this would have been possible is moot; the effort was cut short by the overthrow of the Allende government in September 1973.

And yet, by the very same technology, one could go in wholly different directions. Through the expansion of two-way communication, as in various cable television systems, one could have a complete "plebiscitarian" system whereby referenda on a large variety of issues could be taken through responses back from computer terminals in each house. For some persons this would be "complete" democracy; for others it might mean a more manipulative society, or even the tyranny of the majority, or an increase in the volatility of political discussion and conflict in society.

Without going to either extreme (and, at times, the extremes meet), what is clear is that the revolution in telecommunications allows for a large diversity of cultural expressions and the enhancement of different life styles simply because of the increase in the number of channels available to people. This is happening already in radio, for example, where stations cater to very different tastes, from rock to classical music, from serious talk shows to news and game shows. With the multiplication of television channels and of videocassettes, the variety of choice becomes staggering.

Under free conditions, individuals can create their own modes of communication and their own new communities. No one, for example, foresaw the mushrooming of citizens' band radios and the ways in which they came to be used. These allow strangers to communicate readily with one another. The first, fascinating, social pattern was the development of an informal communication and warning system by truckers on the major roadways, warning one another of traps by law enforcement agents or of road conditions ahead. Sometimes truckers simply used the CB radio to enlarge social ties in what is an essentially lonely occupation.

The CB radio has become a major means of communication in isolated village areas, as in Nova Scotia—a form of community telephone line. And, with two-way video cable television, community interchange may become possible between the elderly or hobby enthusiasts or others with special interests and needs.

In a larger political sense, the extension of networks because of metaphoric face-to-face contact will mean that political units can be reorganized more readily to match, and be responsive to, the scales of the appropriate social unit, from neighborhood to region.

In the end, the question of the relation of technology to liberty is both prosaic and profound. It is prosaic because the technology is primarily a facilitator or a constraint, available to intensify or to enhance —whichever direction a political system chooses to go. It is profound because, as I said, man is a creature capable both of compassion and of murder. Which path is chosen goes back to the long, agonized efforts of civilized communities to find institutional arrangements that can allow individuals to realize their potentials and that respect the integrity of the person.

In short, the question of liberty is, as always, a political consideration. Even speaking of threats to liberty because of the powerful nature of the new surveillance technologies is a misstatement; such a view focuses on technological gadgetry rather than on organizational realities.

Orwell, with his powerful imagination, could conjure up a Big Brother watching all others. But there is not, and cannot be, a single, giant brain that absorbs all information. In most instances, the centralization of such controls simply multiplies bureaucracies, each so cumbersome and jealous of its prerogatives (look at the wars between intelligence agencies in the United States!) as to inhibit, often, the effective use of information.

If anything, the real threat of such technological megalomania lies in the expansion of regulatory agencies whose rising costs and bureaucratic regulations and delays inhibit innovation and change in a society. In the United States, at least, it is not Big Brother, but Slothful Brother, that becomes the problem.

I do not mean to minimize the potential for abuse in our information society. It exists. But there are also agencies of concern, such as the press. Justice William Douglas wrote in 1972: "The press has a privileged position in our constitutional system, not to enable it to make

money, not to set newsmen apart as a favored class, but to bring fulfill-
ment to the people's right to know.''

The crucial issues are access to information and the restriction of any
monopoly on information, subject, under stringent review, to genuine
concerns of national security. The Freedom of Information Act was the
fruit of a long campaign in the 1960s to open up the records of government agencies so that individuals would have access to information
about themselves or information about government agency activities
involving public matters.

In a somewhat different context, when the first large computers were
created, technologists compared them to large generators distributing
energy and assumed that the most efficient model of computer use
would be regulated computer utilities, which would sell computer time
or data services to users. The rapidity of technological change, resulting in the multiplication of mini- and microcomputers, as well as some
second thoughts about the diverse markets for computer usage, led to
complete abandonment of the ideas of computer utilities and recognition of the competitive market as the best framework for computer
development.

The possible growth of the teletext systems described earlier, of
cable television, and of videocassettes may lead to the upheaval of the
major television network systems and to new modes of news presentation, similar to the new competition AT&T faces today in transmission systems.

In sum, from all this arises a moral different from what we might
expect. While technology *is* instrumental, the free and competitive use
of various technologies in an information society is one of the best
means of breaking up monopolies, public and private. And that, too, is
a guarantee of freedom.

Notes

1. See Simon Nora and Alain Minc, *L'Informatisation de la societe* (La
Documentation Francaise, January 1978); and the annual reports of the Harvard University Program on Information Technology, 1976 and 1977.

2. I have expanded on some of these in "Teletext and Technology," *En-
counter* (London), June 1977, and "The Social Framework of the Information
Society," in *The Future of Computers*, ed. by Michael Dertouzos and Joel
Moses (Cambridge, Mass: MIT Press, 1979).

3. For an elaboration of this concept, see my book *The Coming of Post-
Industrial Society* (New York: Basic Books, 1973).

4. See my essay "The Future World Disorders: The Structural Context of

Crises," *Foreign Policy* (U.S.), Summer 1977. For an acute discussion of Rousseau and the problems of size and representation in society, see Bertrand De Jouvenel, *The Art of Conjecture* (New York: Basic Books, 1967).

5. See David Goodman, "Countdown to 1984," *The Futurist*, December 1978.

6. *Science*, November 8, 1978.

Chapter Four

Life in the Information Society

Joseph N. Pelton

If man's five-million-year history were thought of as being spanned by only one month, the following picture would emerge (see Figure 4.1): For 29 days and 22 ½ hours of that 30-day month, man spent his time as a nomadic hunter and gatherer, totally at the mercy of his environment. For 1 ½ hours of this compressed month, man was engaged in agriculture, the harvesting of crops, and building and living in cities. These cities, as at least initially created, were based on trade and agriculture, as well as mutual defense. The last 4 minutes of the month represent the time since the Renaissance, the evolution of the modern nation-state, the emergence of artisans, trade guilds, the middle class, and the gradual elimination of most concepts of serfdom and slavery. This might be thought of as the time at which man entered what Teilhard de Chardin called the "Noosphere" or, in effect, the collective knowledge of scientific man. For "1 ½ minutes," man has lived in the industrial age; and for 50 seconds man has lived in what might be cal-

The views expressed in this chapter are those of the author and do not reflect the views of the INTELSAT Organizations.

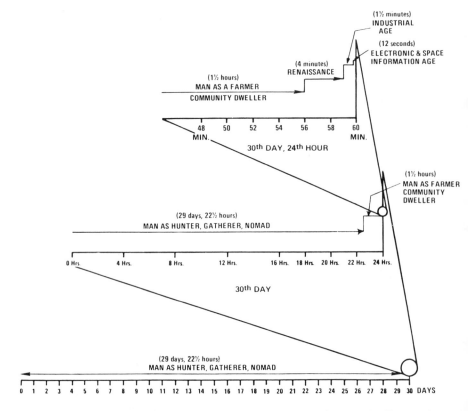

Figure 4.1   History of man depicted as a supermonth. (*Copyright © Joseph Pelton, 1982.   Permission granted*).

led the "electronic information age," if we consider this period as start-ing with the invention of the telephone. Just 12 seconds of the month cover the period of the modern electronic computer and man's ex-ploration of outer space. Who knows what new and exciting changes will come in the next "60 seconds" that separate the information so-ciety from the twenty-second century.

Perhaps another way of viewing the history of "civilized man" is in the progressive development of the size of our cities (see Figure 4.2). Only modern transportation and communication systems have allowed the emergence of the supercity within the last hundred years. For thousands of years cities were held to the scale of pedestrian traffic.

Pretending for a moment that one can see into the future, it can be imagined what might be seen over these hundred years or so, in terms of new technologies and services, new institutions, political develop-

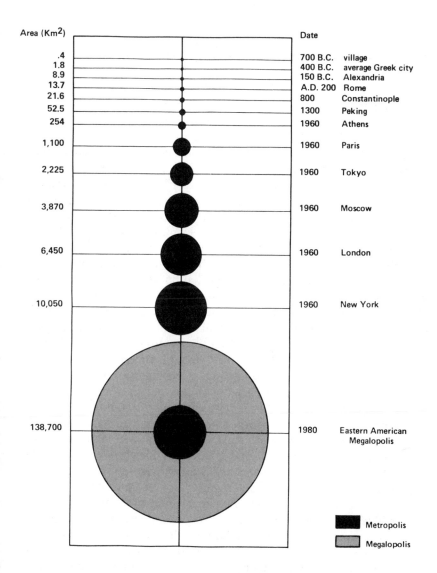

Area (Km²)

	Date	
.4	700 B.C.	village
1.8	400 B.C.	average Greek city
8.9	150 B.C.	Alexandria
13.7	A.D. 200	Rome
21.6	800	Constantinople
52.5	1300	Peking
254	1960	Athens
1,100	1960	Paris
2,225	1960	Tokyo
3,870	1960	Moscow
6,450	1960	London
10,050	1960	New York
138,700	1980	Eastern American Megalopolis

Metropolis

Megalopolis

Figure 4.2 *From CONSTANTINOS A DOXIADIS, "Cities of the Future," Arthur B. Bronwell (ed), Science & Technology in the World of the Future (New York, Wiley Interscience, 1970)* Reprinted by permission.

53

ments, social and economic problems of the future, and new human opportunities. Many needs of the twenty-first century have already been planted in today's laboratories and universities.

Advanced Electronics and Communication

In the field of telecommunications major changes are not unfamiliar. Fiber optics for both long-distance communication and communication directly into the home for data, telephone, and cable TV services is continually being implemented. Equally significant are complex applications satellites now known as "satellite clusters" and "space platforms." These platforms, which weigh many tons and extend for distances in the hundreds and even thousands of meters, provide mobile and fixed communication services; direct-broadcast television; wristwatch telecommunications to firemen, policemen, and other such personnel; and a wide range of earth resource and meteorological applications. The use of robotics, not only in automated factories but in business and even in the home, has become more extensive.[1]

The merging of artificial intelligence, robotics, and advanced telecommunications has dramatically changed the way people live, in terms of schooling, transportation systems, work patterns, manufacturing processes, and so on. The electronic home of the future, the electronic office of the future, and indeed the electronic city of the future, which are characterized in *Global Talk* as the Humanized Telecity, will include such features as teleschooling, telecommuting to work, teleshopping, and teletravel.[2] Integration of new and conventional utilities into a comprehensive "urban macro-architecture" should also become commonplace in the information society. Electronic mail, electronic newspapers, transportation systems, electronic power, and even sewage may all be deliverable to and from the home through a TAGOT network (i.e., the Talking, Going and Thinking network).

Home personal computers will eventually be connected to national and international packet-switched computer networks for a wide range of new telecommunication services. Videoconferencing will eventually find its way into the home of the future; teletravel will also expand in scope, particularly as energy costs continue their upward march. Everything but electronic dating will perhaps become fashionable in information societies.

High-resolution television, three-dimensional television based upon a multiple electronic rastered system and, ultimately, a scaled-down version of holovision, will change at least the data transmission rates to and from the home and office. Unfortunately, the quality of programming, at least in the crystal ball of this author, remains depressingly similar to the Freddie Silverman specials that network television churns out.[3] On the plus side, new telecommunication services should reduce the cost of medical treatment, and telecommunication should help provide effective and affordable schooling for young and old students alike.

The range of services available to the home consumer will be much broader in an information society. The diversity of information, telecommunications services, and entertainment opportunities available will be extremely broad; and the cost of such services will be much less that at persent. Whether the citizen of an information society will be happier or whether these services will increase the quality of life of these citizens, is a point that needs to be considered.

New Institutions

In what follows three different kinds of institutions—social and cultural, business, and government—are examined. Global communication, especially global television, can be expected to have a considerable impact on social and cultural institutions and practices. Morris Desmond, the pop anthropologist, in particular has noted that communication satellites have enormous influence in the direction of what he calls "supertribalizations." The ability to have ready access to information about the rest of the world cannot help but have an impact on local belief systems, cultures, and religions. Only concerted efforts to preserve local customs and traditions can be expected to succeed against the almost irresistible trend toward global homogeneity, in terms of popular taste in art and culture.

Robert Hughes, the Australian art critic, claims that electronic mass media and the McLuhan concept that the media is the message is "turning our brains into corn flakes." Hughes argues that electronic media provide information many times that of previous generations and that because this maelstorm of information races through our heads and cannot be processed, people are simply left with pattern recognition. Repetition and size, particularly gigantism, become the value dimension, not quality. The rush of information also seems to create a preoccupation with the present. Historical perception or future goals are seldom well handled by electronic media. Network television news in the United States has been castigated as the "cheerleader on the side of today's victors." The threat to grass-roots social and cultural institutions, to historical tradition and localism, seems unmistakable in such a world.

There are those who suggest that telecommunication will dramatically reshape commercial institutions as well. Hewett Crane, in his book *The New Social Marketplace*, has suggested that the commercial sector will increasingly enter into the area of social services, in terms of education, cultural activities, and so on, to create a counterpart to the consumer marketing network.[4] Alvin Toffler, in his book *The Third Wave*, boldly, skillfully, and implausibly suggests that robotics, artificial intelligence, and sophisticated new computer software will allow businesses to specialize, decentralize, and "respond effectively to individual consumer needs."[5]

Both Toffler's and Crane's conceptions are interesting and, in many ways, "desirable," but there is little objective evidence to suggest that information societies will see changes designed to preserve localism, diversity, and responsiveness to individual needs within the commercial marketplace. Indeed, there is a good deal of evidence that the business world of the future will change in the opposite direction. In particular, economies of scope and economies of scale will emerge from the marriage of previously disparate activities of telecommunication,[6] transportation, and energy. Certain technical and economic trends seem to be driving the twenty-first-century business world toward both vertical and horizontal integration.

These superenterprises will be interconnected largely through the need to maintain large research laboratories, extensive marketing and distribution systems, high levels of competence in materials research, and robotics. Solid-state technology electronics and photonics will likely dominate the business world of the future. One of these prototypal superenterprises is characterized as the "telecomputerenergetics" industry.[7] The Exxon Corporation already begins to resemble a prototype of such a corporation in that it is not only heavily involved in petrochemical energy production and distribution but also is involved in solar energy, geothermal energy, electronic office systems, telecommunications systems, and robotics. Hughes Aircraft Company and TRW also span these technologies, as well as aerospace technology. Exxon is also developing a line of electric vehicles as a logical consequence of its purchase of Reliance Electric. Thus one can foresee the emergence of supercorporations, involved in technologically interrelated enterprises, in highly developed national economies such as the United States, Japan, and Germany. Indications are that Germany's Volkswagen and MBB are headed in the same direction, thus perhaps showing an emerging global pattern.

Alvin Toffler has noted with great clarity the characteristics that mark the modern corporation in terms of centralization, standardiz- ation, synchronization, and so on.[8] These factors, coupled with the eco- nomies of scope that one would anticipate in superindustries like the telecomputerenergetics industry, will create enormous pressures for consolidation of activity and a bias in terms of standardization, not only of quality, but also of management techniques and business philosophy. Although artificial intelligence and advanced computer capabilities present opportunities for decentralization and greater responsiveness to individual needs, this does not mean it will happen.[9]

Global inflationary trends, coupled with economies of scope, econ- omies of scale, and the pursuit of greater productivity and cost efficien- cy, can be expected to create strong counterpressures toward homogeneity in the business marketplace in an information society. Thus, while one may find superficial diversity increased rather than diminished, commercial integration seems likely to be the predominant trend of the future. Megabusinesses rather than electronic cottage in- dustries are likely to be the wave of the future.

How will this affect governmental institutions? Regardless of the political philosophy employed in an advanced industrial society, be it communism, socialism, or free market enterprise systems, some of the key "economic premises" are similar. Most modern governments base their economic programs on increased economic throughput. Consum- ers may be happier and freer in a democratic state than under commun- ism, but economic throughput is essential to the survival of communis- tic and democratic states alike. The most likely view of the information society from an economic perspective, therefore, is one wherein gov- ernments mimic the innovations introduced in commercial and indust- rial fields; further centralization, standardization, and attempts at syn- chronization of programs at the national level can be expected.

Environmentalists and organizations such as the Club of Rome have indicated that our approach to economic growth, resource exploitation, and the like cannot be continued indefinitely without creating enor- mous environmental damage and the exhaustion of readily available and exploitable resources. Comparative studies of the electronics and information industries in France, the Federal Republic of Germany, Japan, the United States, the United Kingdom, and the USSR tend to confirm the fact that patterns of economic growth and development all follow similar treads in terms of aggregation and integration of large economic enterprises. Although certain governments have encouraged such integration through legislative mandate and mandatory merger, other societies have essentially created the framework for this to occur by removing barriers to competition among giants and weaklings.

When even liberal economists, such as Lester Thurow of MIT, start advocating the elimination of antitrust in the United States and the logic of there being an evolution toward a single Detroit car manufacturer, the trend seems clear. Despite romantic storytelling to the contrary, Goliaths usually win and Davids get trounced in the modern international business world. International economic competition has daily accelerated the pace of consolidation of industrial and commercial activities. The growth of governmental agencies, civil service employment, and taxation have also shown that government bureaucracy follows the business trend of bigger being better.

The Social Challenge of the World of Global Talk and Global Think

Emerging telecommunication systems, as well as new institutions shaped by these technologies, will obviously give rise to many social, economic, and political problems. These problems include potential invasion of privacy, information overload, loss of individual identity, increased opportunity for totalitarian control of the individual by government, technological unemployment, and such sociopsychological problems as readjusting to new and potentially disturbing man/machine relationships. Each of these problems should be examined in more detail.

Privacy

There have been a number of commissions created in Western countries to examine the problem of the invasion of privacy. These commissions, such as the Younger Commission in the United Kingdom and the U.S. National Commission for the Review of Federal Laws Relating to Wiretapping and Electronic Surveillance, have generally concluded that the problem of privacy invasion exists to varying degrees.[10] Each of these national commissions has recommended a variety of approaches toward providing the individual with greater privacy, but none of the approaches has proven successful in stemming the trend toward more rather than less invasion of privacy.[11]

There has also been increasing concern about invasion of privacy at a macro level—whether large, sophisticated, computerized systems operating from the United States, Japan, or France, for instance, might be able to compile information about nations and their economies and exploit it to the disadvantage of a smaller country that does not possess comparable capabilities. This concern has been manifested in such areas as the improper use of earth resource satellite data and the uses to which credit card companies put data compiled on overseas cardholders.

Among the solutions suggested to deal with privacy issues in the information era are digital encryption techniques to better secure the privacy of information to authorized users, as well as guaranteeing rights of access to individuals to inspect data files compiled on them, in order to correct or challenge the accuracy of information. In a Delphic Survey,[12] a majority of experts believed that nearly *universal data bases* would be compiled on "all" people in information societies by approximately the year 2000. Everyone would suddenly be in *Who's Who*.

Information Overload

Telecommunication systems are generating information at an exponentially increasing growth-rate curve. The amount of information used by humans, however, is likely to increase at a very slow and marginal rate (see Figure 4.3). Certainly humans are able to absorb a smaller percent of the increased information becoming available each year.[13] Professor Yasumasa Tanaka in a recent study concluded that information in Japan was growing at a rate significantly greater than Japanese citizens were able to assimilate and use the information. The truth is that much of the new information being generated and transmitted through the global telenet is processed and analyzed by machines, not people. Indeed, a prime function of the future librarian and the personal computer will be to screen and filter information, to present only the most relevant information to those seeking to know about a particular subject.

To help analyze the rates at which people use information, the following terms will be used—TIUPIL and GHIUD.[14] TIUPIL stands for Typical Information Use per Individual Lifetime, which is "assumed" to be approximately 650 million words, or the equivalent of some 20 billion bits of linguistic information that find their way through the "average" human cortex in the course of a lifetime. GHIUD stands for Global Human Information Use per Decade. With a global population of 4 billion, a GHIUD is equivalent to some 400 quadrillion words or some 12 quintillion bits of information. This represents the totality of human verbal and written intercourse used in a given decade.

Both advanced communication satellites and fiber optic cables are capable of transmitting a billion bits of information per second. Put another way, the entire *Encyclopedia Britannica* could be sent through a telecommunication system six times a minute. Translated into

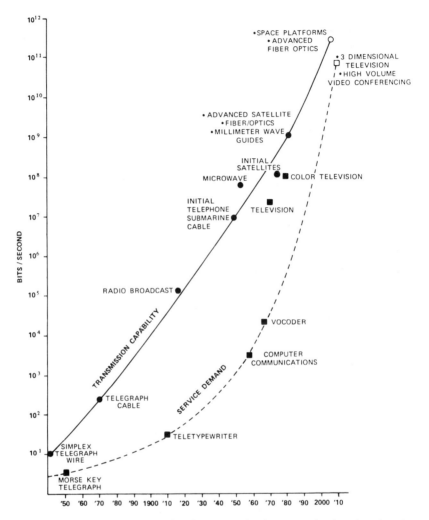

Figure 4.3 150-year look at development of telecommunication (service demand vs. transmission capability). (*Copyright © Joseph Pelton, 1982. Permission granted*).

TIUPILs, these two systems could handle 200 TIUPILs an hour, or some 4,800 TIUPILs a day. A network of six advanced communication satellites could thus polish off the collective lifetime thoughts and speech of all New Yorkers in less than a year. Twenty-first-century space platforms that have communication-carrying capabilities of 100 billion bits per second are already well-defined concepts. Two such platforms would be capable of handling in a day the entire U.S. population's thoughts and writings for a decade. Why are such large communication systems conceived? In an information society these systems will carry not only telephone, telex, and telegraph messages but also television, videoconferencing, and perhaps high-resolution television or even three-dimensional television. The transmission of a rapid sequence of holographic images, for instance, would require a bit rate of 1 trillion bits of information per second. Such advanced telecommunication systems are thus oriented toward machine-to-machine communication.[15]

More significant is the potential of hundreds of newspapers, research articles, books, visual images, and the like being spewed forth from on-line computer terminals into the home. Sociologists, Robert Merton and Paul Lazarsfeld in the 1930s did extensive studies of animal behavior when exposed to extremely high levels of information content. Over a period of time, the results showed that the research animals became extremely apathetic, lost their appetites for food, and as information was increased, lost their sexual drive and eventually died. Tanaka, in studying the information overload phenomenon, concluded that a commercial organization seeking to win approval for controversial projects over citizens' opposition (e.g., in the case of a nuclear power plant) might be most successful by simply overloading the affected community with information. Effective political action and organization might become impossible in the face of being bombarded with a maelstrom of extraneous information.[16]

Telecommunication as an Instrument of Suppression, Terrorism, and Extralegal Activity

What about the use of electronic technology for political control, suppression, or extralegal activities? Many of the technologies envisioned by George Orwell in *1984* are possible now, and many more will be available in the information society. In France, all employees of a particular plant wear electronic badges that can be monitored by an IBM computer. Although this is done for industrial health protection, the labor unions have objected to employee surveillance. By the year 2000 it is likely that space platforms will be built to provide two-way wristwatch radio communication to assist police, firefighters, and security personnel in their work. This "Dick Tracy" aid, however, could be used for more sinister purposes, such as becoming a permanent handcuff monitor for checking on the movements of political subversives. Two-way television, computer monitoring techniques (such as keyword analysis), and satellite surveillance could be used to achieve degrees of political control and suppression undreamed of in former times.

In addition to the problem of government use of telecommunication is the specter of electronic criminals and electronic terrorists. In a highly centralized computer- and communication-dependent world, technologically astute criminals, spys, and terrorists have the potential to inflict everything from mischief to catastrophe. Embezzlement, credit card and magnetic card forgeries, destruction of vital records, covert acquisition of secrets, falsification of data to incriminate or exonerate key officials, destruction or immobilization of transportation systems, weapon systems, communication systems, credit and banking systems—indeed any system dependent upon telecommunication—are potentially vulnerable to an adept saboteur. It is conceivable that a computer programmer/systems analyst with top-secret information and the right equipment could engineer a nuclear war between the United States and the Soviet Union.

Other Problems

While preinformation societies may marvel at the prospect of an emerging telecommunication network, they will also likely experience future shock. Religious, cultural, and social beliefs will be challenged. Machines that are smarter, quicker, more reliable, and perhaps ultimately more innovative than the typical citizen may well emasculate self-esteem, create violent and destructive emotions, and give rise to a twenty-first-century version of Dadaism or nihilism.

Skcepticism in regard to technological progress is also commonplace. Predictions as to social impacts, however, are quite diverse. Marshall McLuhan said: "At the speed of light people lose their goals in life." He further observed that "since Sputnik there is no Nature. Nature is an item contained in a man-made environment of satellites and information."[17] The threat to the individual's self-image means, in McLuhan's view, that mass ideologies and even mass religiosity may hold the key to the twenty-first century:

> Psychic communal integration made possible at last by the electronic media create the universality of consciousness foreseen by Dante when he predicted that man would continue as no more than broken fragments until they were unified into an inclusive consciousness.[18]

Certainly the force of satellite communication to create a global cohesiveness is awesome. Billion-person television audiences will become commonplace. The individual and individualism will have to struggle for survival against the "intelligent global machinery" of the twenty-first century.

Opportunities of the Electronic Future

Fears about the information society are great, but the litmus test in regard to social problems is simply this: "Who, given a choice, would go back?" Who would truly return to horseback and handbills over telephones and telexes? Jets, telephones, and electronic mail are somehow more convenient. Air conditioning vs. serfdom; a tropical vacation vs. slavery; a nine-to-five job vs. black-lung disease; allergies vs. the Black Death.

There is little doubt that modern society cannot function without the extensive telecommunication systems that tie societies together. If the telecommunication systems of an information society were suddenly turned off, that society and others tied into its information systems would come to a standstill. Each year $5 trillion of money is transferred overseas in the Eurodollar, the Asiadollar, and similar currency markets. Reservations for worldwide airline systems scheduling thousands of trips per day are also dependent upon these systems. World trade, in its current form, cannot exist without these systems. Much the same pattern holds for national economic systems.[19]

The truth is that without modern technology, telephones, vaccines, computers, jet airliners, and automated industries, the modern global population of some 4 billion people simply could not exist. We would die of starvation and disease, or if we could survive, our standard of living would be far lower. The issue, then, is not one of declaring a technological moratorium or going backward; the issue is in recognizing that there is no such thing as a perfect world and that new social problems will always substitute for old ones. In effect, these problems must be minimized as we maximize new opportunities. Certainly new opportunities exist. They exist in exciting, exotic, even mind-boggling forms.

Telecommunications is an extremely powerful tool for economic development. Telephones; in particular, have a tremendous economic multiplier effect in terms of generating new investments elsewhere in the economy (Figure 4.4). These impacts, in terms of dollars, are up to four times greater than the initial investment in telecommunication. If one compares the historical development of the United States over the last two hundred years (Figure 4.5), as measured in employment patterns, with a snapshot comparing the various countries of today's world (Figure 4.6), the similarities in the graphs are startling. One of the best indexes of economic development yet developed is a correlation between telephones per capita and economic prosperity (Figure 4.7). The correlation is well above 80 percent, with telephone investment typical-

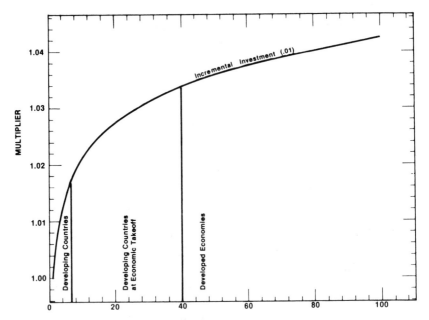

Figure 4.4 Telephones per 100 population. (*From Andrew P Hardy, "The Role of the Telephone in Economic Development," Institute for Communication Research, Stanford University, January 1980*). *Reprinted by permission.*

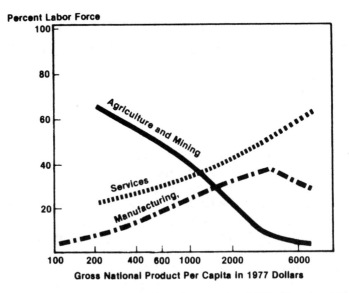

Figure 4.5 World labor as a function of per capita GNP. (*Reprinted by permission of Oxford University Press*).

64

Percent Labor Force

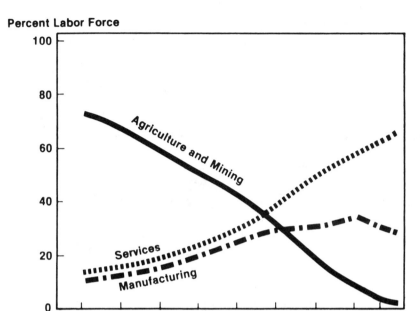

Figure 4.6 U.S. labor composition, 1820–1970. (*Reprinted by permission of Oxford University Press.*)

ly leading (in a chronological sense) economic growth. Today, low-cost earth terminals in the range of $20,000 to $50,000 per terminal are being developed to operate with INTELSAT satellites. These new terminals, together with newer and higher-power satellites, should be able to provide rural educational radio and television, as well as two-way data and telephone service into the most remote parts of the world, and to do so at reasonable costs. There are perhaps few instruments of social change and development more powerful and with greater potential than the emerging telecommunications networks.

Certainly, the fundamental components of an information society— modern telecommunication, robotics, and computers—are far from a panacea. They bring with them enormous political, economic, and social problems. Yet one cannot hide the fact that they also bring enormous opportunities—to escape global starvation, to diminish illiteracy, and even to aid the cause of world peace. The world of global talk and the world of global think must be thought of as unexplored worlds in which we journey to the future. Dangers, as well as opportunities, are enormous. Yet, if the time is taken to analyze the nature of the problems and opportunities inherent in an information society and plan with care, information societies can succeed. Societies must seek to

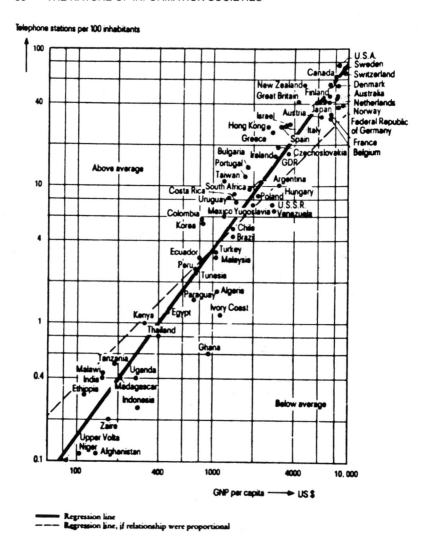

Figure 4.7   Correlation of telephone stations and GNP; status as of January 1, 1978. (*Reprinted by permission of* Siemens Technical Review.)

temper cost efficiency with cultural integrity and to temper economic throughput with the higher objectives of survival and humanistic sensitivities. These are the keys to making the future better than the past. Humans may yet establish new forms of freedom by colonizing the solar system, may travel to other star systems, and may even create a new sun to power societies of future generations. But Anthony Oettinger's words must be recalled: "Man must choose his future steps very wisely if he is to avoid dancing the dance of dinosaurs."

Notes

1. For a detailed account of this new industry and the research being done, see Pamela McCorduck, *Machines Who Think* (San Francisco: W. H. Freeman, 1979).

2. Joseph Pelton, *Global Talk* (Alphen aan den Rijn, Netherlands: Sijthoff & Noordhoff, 1981).

3. Robben Fleming, president of the Corporation for Public Broadcasting, made a similar assessment in a paper presented at the 1981 Edward R. Murrow Symposium, Pullman, Washington.

4. Hewitt Crane, *The New Social Marketplace* (Norwood, N.J.: Ablex, 1980).

5. Alvin Toffler, *The Third Way* (New York: Knopf, 1980).

6. The term telecommunications, as used here, includes computers and electronics.

7. In Pelton, *Global Talk*.

8. Toffler, *The Third Wave*.

9. One must not misjudge recent trends. Distributed processing in national computer networks is no more of a decentralizing phenomenon than are McDonald's hamburger franchises.

10. One can, for example, through the right contact buy for about $500 a copy of any personal FBI confidential file.

11. The primary concern of these commissions is wiretapping and unauthorized access to confidential computer files.

12. Conducted among some 80 telecommunications and aerospace experts. Joseph Pelton, "The Future of Telecommunications," *Journal of Communications* 31, no. 1 (Winter 1981): 177–89.

13. Alvin Toffler, *Future Shock* (New York: Random House, 1970).

14. These terms were first used in Pelton, *Global Talk*.

15. Although individual viewers might watch a rapid sequence of holographic images for entertainment purposes, the viewer would perceive only broad forms of pattern recognition. Far less than 1 percent of the total information content could ever be assimilated by a human brain. Although a human would perhaps be able to use an extremely small fraction of the information content of a holovision transmission, this would at least be a relatively benign environment for the viewer.

16. An informal experiment, designed to determine the percentage of magazines, journals, and reports entering a typical home that were actually read, concluded the percentage was less than 5 percent

17. Cited in Stewart Brand, "McLuhan's Last Words," *New Scientist*, January 29, 1981, p. 296.

18. Ibid.

19. Whether the issue is rice commodities in Indonesia, travel agencies in Kenya, a credit card corporation in the United States, or a manufacturing organization in France, all are dependent on modern computers and modern telecommunication systems for paying bills, maintaining inventories, providing employee benefits, ordering supplies, or maintaining accounts. The dependence on information is there.

PART 2

PUBLIC POLICY FOR AN INFORMATION SOCIETY

Chapter Five

Public Policy for the Information Society: The Japanese Model

Ryuzo Ogasawara
and
Jerry L. Salvaggio

In Part 1 of this book Daniel Bell, Herbert Schiller, Ernest Rose, and Joseph Pelton called attention to what they consider to be the most significant cultural and social implications of living and working in an information society. They cited possible threats to individual liberty, the centralization of information technology, the invasion of privacy, economic inequity, and the potential for information monopoly. In addition, Tony Rimmer, William Read, Roland Homet, and Jennifer Slack make it clear in this part of the book that existing U.S. policy planning, rules, regulations, and technology assessment regarding telecommunications are not up to the task of protecting the public from the adverse effects associated with a wired society. Rather, telecommunications in the United States seems to be developing solely on a profit motive basis without the benefit of a long-range public policy.

If telecommunication systems and services are not to "run amok," to use Schiller's phrase, a carefully designed public policy needs to evolve in the United States. Telecommunication systems will form an economic base for information societies in the same way that that the telephone industry forms an economic base in the industrial society. Mass public use of telephone service could not have resulted without a long-range plan for its use. Thus it is logical to conclude that instead of allowing telecommunication systems to develop on an ad hoc basis, a determination needs to be made of how they can best serve to solve existing socioeconomic problems and be used constructively in the information society.

Several countries have moved far more rapidly than has the United States in preparing for the information society. England's Prestel, Canada's Teledon, and the French Antiope systems are but a few examples of national videotex systems that have evolved in other countries as a result of specific goals and the consideration of future information needs.

As an alternative to allowing telecommunications to be developed by the private sector alone, without the guidance of a comprehensive public policy, the United States might do well to follow the model of Japan, a country that has moved enthusiastically toward becoming an information society since 1965.[1] Japan's government has been heavily involved in researching the nature of information, has invested billions of yen into experimental information systems, and has a long-range plan for replacing traditional media with advanced forms of telecommunications. The goal is to utilize the potential of the technology to meet all of Japan's future information needs in the fields of education, journalism, business, and government. Put in another way, Japan is preparing for the information society in a manner that may minimize many of the adverse effects that the authors in this book have addressed.

Not only is there much to learn from the Japanese in regard to preparing for the information society, but it may be educational to observe the consequences of a necessarily close alliance of government, the information industry, and public utilities. Rose's and Schiller's fear of the control of technology and monopoly of information could be a consequence of the close affiliation of the government and the information industry.

Before examining Japan's plans for the information society, it must be understood that the infrastructure of Japan's regulatory body in regard to telecommunications is considerably different from that of the United States. Though communication is not government operated or controlled in Japan, the Ministry of Post and Telecommunications (MPT) is delegated with the responsibility of regulating not only broadcasting but telephone and postal services as well.[2] Along with this onerous responsibility, the MPT has a hefty budget for research and development. This has never been the case with the Federal Communications Commission (FCC), which is inadequately funded. This places the MPT in the enviable position of being able to design comprehensive plans for integrating all telecommunication systems for the benefit of the public.

In order to design a system that will best serve the public in an information society, an assessment of the growth of information flow is needed. Since 1965, the MPT has been conducting a survey of Japan's consumption of information. The survey is designed to provide a quantitative analysis of the annual flow of information.[3] Thirty-four kinds of media are covered including computer communication, books, broadcasting, personal conversation, and mail. Each medium is considered in terms of (1) how many words are carried, (2) how far, and (3) at what expenses.

The MPT census makes a clear distinction between "traditional media" (personal conversation, theatre, education, etc.) and "telecommunications media" (computers, broadcasting, satellite signals, etc.). Perhaps the most significant finding is that telecommunications media are economically more effective than traditional media. While personal media channels barely amount to 1 percent of the supply and 23 percent of the consumption, mass media channels account for 99 percent of the supply and 77 percent of the consumption.[4]

The information census has made it clear that technology is changing relationships among media. Traditional forms of media face particular problems, while telecommunications media are becoming more efficient. In terms of information flow, the amount spent in direct contact communication media is greatest and occupies 69.4 percent of the costs for all media, whereas the amount spent in telecommunications is only 23.8 percent.[5] What is significant is that while the volume of information flow through direct contact communication media remains constant, the unit cost is becoming more and more expensive. On the other hand, the unit cost of telecommunications drops visibly in spite of the increasing volume of information flow. Put another way, traditional media (mail, telephone, newspaper, and TV broadcast through terrestrial networks) seems to have reached their full potential. The development of telecommunications (computer communication, facsimile transmission, communication satellites, videocassettes, videodiscs, videotex, teletex, etc.) is quickly being introduced and shows rapid rates of diffussion. By 1970, the MPT census made it clear that new forms of electronic media are far more efficient and offer a 'greater variety of information services than traditional media do. Furthermore, the potential of telecommunications is unlimited because the ability exists to transmit a voluminous mass of information at high speed and at relatively little cost.

Based on the above research conclusions, Japan instituted a number of experimental telecommunication systems including direct broadcast satellite, fiber optics, coaxial cable, and data facsimile terminals. Each experiment was designed by the MPT as a means to learn more about the public use of information.

Tama New Town was one of the first experiments. Through a coaxial cable information system (CCIS) three hundred apartments were interconnected with various forms of communication technology. The project was a joint experiment of the Nippon Telephone and Telegraph Corporation (NTT), the Living-Visual Information System Development Association, and the Ministry of Post and Telecommunications. The experiment, begun as early as 1972, included pay TV, facsimile newspaper service, an electronic memo-copy service between individuals, still picture requests, and a two-way subscriber keyboard terminal.[6]

Following the lead of England's Prestel, Japan has also been experimenting with viewdata. The MPT and NTT began a co-sponsored project known as CAPTAINS (Character and Pattern Telephone Access Information Network System) in 1979. Information providers include newspapers, news agencies, TV stations, and smaller organizations. The MPT has done extensive research on public attitudes toward CAPTAINS and expects the service to be popular when terminals become less expensive.

Realizing that the cost of accessing data on CAPTAINS will be too expensive for everyday use, the MPT has also invested in a one-way form of video communication. The experiment is called TELETEXT. The NHK (Nippon Hosa Bunka) has been conducting tests on consumer use of the system since 1978. TELETEXT might be used for up-to-the-minute news, airline schedules, and headlines; whereas CAPTAINS might be used for accessing more in-depth information and exchanging electronic memos.

Anticipating the future, Japan wired an entire community with fiber optics at a cost of $16 million. This extremely sophisticated telecommunications experiment, conducted between 1978 and 1981, foreshadows the day when telephone wires will be replaced by fiber optic cables in all information societies. Known as Hi-Ovis (Highly Interactive Optical Visual Information System), the experiment was located in Ikoma, outside Tokyo, and involved 158 homes and 10 public institutions.[7] In addition to videotex, Hi-Ovis offered security alarms, retrieval of still pictures, a videoconferencing link to NHK studios, remote playing of videocassettes from the control center, and local program origination. The most exciting aspect of the experiment was clearly the two-way visual component. Participants had a small TV camera installed on top of their TV sets that allowed them to visually and aurally join locally originated programs in progress.

Perhaps the best example of government, industry, and public utilities working together is direct satellite broadcasting. While the government has been experimenting with direct satellite broadcasting, Nippon Telephone and Telegraph, in collaboration with six electronic firms, has developed Subscription Fax, an inexpensive home facsimile terminal that will be able to receive news and information directly from satellites. The *Asahi-Shimbun* and other newspapers and information sources will provide the content. Together the three elements—direct broadcast satellites, home facsimile terminals, and information providers—will allow for a nationwide electronic news delivery system at a minimal cost to the user.

The manner in which government, industry, and public utilities have worked together to develop experimental telecommunication systems indicates that Japan will be one of the first (if not the first) information society. What will the social impact be? While this remains largely unknown, it is suggested here that the building of large experimental telecommunication systems, such as those described above, the conducting of research on public use, and the development of long-range plans may lead to fewer adverse effects than allowing telecommunications to evolve strictly on a commercial basis. An analogy might be transportation: Japan's national railroad system offers a far greater service to the public than does the U.S. subsidized railroad system. Information inequity, for example, should not be a problem in Japan because direct broadcast satellites will cover the entire nation and the NTT is developing inexpensive facsimile terminals. In the United States, by contrast, teletext and videotex are being developed privately by AT&T, cablecasters, broadcast networks, and the newspaper industry. Green Thumb, an experimental information system funded by the Department of Agriculture and the Department of Commerce, is an exception. Thus far, the United States has not shown any interest in designing or funding anything similar to CAPTAINS or TELETEXT. The outcome of a lack of central policy planning may be a potpourri of dissimilar systems available only to those fortunate enough to live in areas that have been wired. Even in these areas only those affluent enough to purchase the systems and access to the information will be able to enjoy the benefits.

It can be seen that Japan is approaching the information era with the same zeal with which she approached the postwar industrial period. What lessons are there to be learned from Japan's precocious recognition of the value of an information society and her early attempts to prepare for this new era?

First, Japan has one government agency, the MPT, entrusted with the task of preparing for the information society. In the United States a number of governmental agencies are involved with the development or regulation of telecommunications including the National Telecommunications and Information Administration, the FCC, the ICA, and the Office of Technology Assessment; but none has the budget or the authority of the MPT.

A second lesson to be learned is that there is much to be gained when government works in conjunction with, rather than against, the private sector. While the Justice Department in the United States, throughout the 1970s, was attempting to break up AT&T and IBM because they had grown too large, the Ministry of Post and Telecommunications in Japan was working with Nippon Telephone and Telegraph and a dozen electronic firms and newspapers to build the telecommunication systems that will form the basis of the information society. The cost of building fiber optic networks and direct broadcast satellite systems capable of transmitting a national university over the air to home facsimile terminals is sufficient enough reason to encourage such an alliance. Working together, the government, public utilities, and the information industry are able to develop telecommunication systems that will be commercially viable and at the same time will be designed to meet the needs of the public.

A third lesson that can be taken from the MPT research is that telecommunication is more cost efficient than traditional media. This suggests that a country's economy will benefit from a gradual transformation from the latter to the former. Part of Japan's economic success as an industrial society can be attributed to early modernization of industrial plants. In the same way, building new telecommunication systems will benefit the Japanese economy in the information era.

What remains to be seen are the social, cultural, and political consequences of allowing the triumvirate of government, public utilities, and the information industry to form the infrastructure of the information society. This places the most important technology of the future in the hands of a small but powerful group. And with the centralization of technology comes the potential for controlling information.

Already in Japan ownership of the largest information systems are limited to a half dozen large corporations. The five leading newspapers (whose combined share amounts to almost 50 percent of the total circulation in the nation) also control many TV and radio stations. Their influence on public opinion cannot be overestimated.

Literature has shown that information is the measure of power today and will be more so in the information society. The centralization of information systems could deter the expression of a diversity of issues and viewpoints which are crucial to democratic societies. On the other hand, an alliance of government, public utilities, and the information industry is certainly preferable to control by government alone. The close alliance should result in a healthy countercheck system between government and industry.

In summary, Japan's preparation for the information society demands attention if only for the economic stability it will bring. In all likelihood, planning for the information society will bring more than prosperity to Japan. It will minimize the adverse effects to some degree. At the least, Japan now has in place an infrastructure for dealing with the problems, and by virtue of having acknowledged the fact that they will be an information society, Japan has taken a significant step toward meeting the challenge of the telecommunications revolution.

Notes

1. Since 1965, Japan's Ministry of Post and Telecommunications, comparable to the FCC in the United States, has collected statistical data on the flow of information with the knowledge that modern nations were rapidly becoming information societies. For more on Japan's early recognition of information societies, see Alex Edelstein, John E. Bowes, and Sheldon M. Harsel, *Information Societies: Comparing the Japanese and American Experience* (Seattle: University of Washington Press, 1978).

2. All statistical data presented here have been extracted from the Ministry of Post and Telecommunications. See *Report on Present State of Communications in Japan*, Fiscal 1980, edited and published by The Look, Japan, supervised by the Ministry of Post and Telecommunications (n.d.).

3. The MPT survey does not consider the qualitative (value-pertaining) aspects of information.

4. For a more detailed analysis of MPT objectives and methods of measuring information, see Jerry L. Salvaggio, "An Assessment of Japan as an Information Society in the 1980's" (presented at the 1982 World Future Society Conference, Washington, D.C.)

5. See *Report on Present State of Communications in Japan*, p. 10, for two charts comparing telecommunication and personal media costs.

6. A detailed analysis of the Tama New Town experiment can be found in Visual Information Development Association, *Report on TAMA CCIS Experiment Project in Japan: 1978* (Tokyo, Japan: n.d.). For an excellent summary of the experiment see Yoneji Masuda, *The Information Society as Post-Industrial Society* (Tokyo: Institute for the Information Society, 1981).

7. Hi-OVIS Project: Hardware/Software Experiments, July 1978-March 1979, *Interim Report by the Visual Information System Development Association-MITI Juridical Foundation* (Tokyo, Japan: 1979).

Chapter 6

Monopoly Versus Competition: Social Effects of Media Convergence

Roland S. Homet, Jr.

It is undeniable that with the newly emerging telecommunication systems and services, market competition has replaced regulated monopoly as the governing force in the United States. That is not to say that monopoly or oligopoly has entirely disappeared or even lost importance. We have but to think of the broadcasting networks' continuing control over prime-time television or the exclusivity of local telephone service. But the more dynamic developments, such as cable, direct satellite transmission, data processing, and videotex, have not been assimilated by these earlier and established monopolies. Instead they have been treated, after a period of uncertainty, as service offerings in their own right with no prescriptive domination by telephone companies, broadcasters, newspapers, or the like. We all know the mileposts, from *Carterphone* to cable deregulation to *Computer II* and the rewrite legislation of today. The process of opening up new markets to new entrants has spanned administrations of both political parties and seems unlikely to abate as we move into the information era.

The result is to usher in an era of what is sometimes called "media convergence." The market entry of a new firm depends less on the traditional categorization of that firm and more on its technical and economic competence, and most important, its entrepreneurial initiative. To be sure there is resistance to this blurring of boundaries, but increasingly newspaper companies and telephone and broadcasting companies are coming to see themselves as *communication enterprises*, on the same footing for this purpose as new apparitions like Radio Shack or Exxon Information Services or British Telecommunications. Any one of these can offer data processing, computer-controlled information responses (videotex), direct satellite connections, and other new services; perhaps more significantly, none can resort to regulatory protection to block any other from doing so.

But that is of course a prescription for territorial struggle in the keenest degree—particularly when one considers that videotex can shift advertising and financial tabulations away from newspapers and that direct satellite connections can bypass entirely the local telephone loop and the local broadcasting affiliate. As regulatory intervention fades, who or what will preside over the outcome? Will the struggle be left simply to the survival of the fittest, or as Tennyson put it, to "nature red in tooth and claw"? And where will be our social values then?

It may tell a lot that the major structural issues of an information society, the determination of who gets to offer what services under what terms and conditions, seem likely to be decided not by Congress and not by the Federal Communications Commission and not even by the courts, but by a process of sealed-off, bilateral bargaining between the two largest actors in the field—AT&T and IBM—and the Department of Justice. In the future it could well be that antitrust will become a more significant policeman of market entry and market behavior than this country's communication laws. It bears remembering that in 1980 it was the Antitrust Subcommittee in the House of Representatives that refused to permit a vote on the common carrier regulatory reform legislation produced after five years of effort by its sister Subcommittee on Communications.

If that implication is carried forward, if private bargaining with the government comes to replace public decision making, then we are going to have to think carefully about the ramifications for democratic participation and control in the information society. Even with the opportunity for comment by interested parties on a draft consent decree, antitrust settlement negotiations are essentially private affairs with no meaningful public record or opportunity for judicial or legislative review. The 1956 AT&T decree has been with us for more than twenty-five years and it has almost no defenders. The possibilities of abuse or of overreaching by either side are there. So also are the possibilities of an inadequately considered policy judgment. As one who, in an earlier White House, championed the virtues of market competition for communication services, I remain committed to the merits of innovation and efficiency and customer responsiveness and rational pricing. But there are clearly noneconomic factors to be considered as well, what my economist friends call "externalities," but what I prefer to call the "values dimension."

A listing of the pertinent social values might well include the following:

1. Personal control over the communication stream
2. Protection of personal privacy
3. Shared communal experience
4. Equality of access and participation
5. Cross-cultural understanding
6. Freedom of expression
7. Creative opportunity
8. Responsiveness to consumer needs
9. Stability of service offerings

One might easily think of others, but these are enough to test against the new communication marketplace.

Personal control is one value that seems clearly strengthened by the new technologies and services. Tired of "Love Boat"? Put on a videodisc of your own choosing or use a cassette to record something you missed or invest in a pay cable program. Nerves jangled by a telephone call when you're trying to think? Store-and-forward technology can spare you the disruption until you're ready to receive. Many of the new services are interactive—electronic mail, computer interrogation—and videotex turn the passive recipient of messages into an active participant. I personally believe this is one of the most highly prized and socially meritorious attributes of the new communication dispensation. But it has not yet wholly prevailed. Witness the infuriatingly anonymous errors of computer billing.

Protection of privacy has commanded a good deal of policy attention as computers have come into greater use. Much of the new communication capacity is suspect: computer memories, cable scanning of viewer preferences, the trail left by electronic mail. Interestingly enough, two technologies that are immune to economic regulation—videodiscs and cassettes—are also free of privacy infringement potential. But bank deposits and business records are vulnerable to unauthorized capture, and a political dictator could order electronic seizure of

all sorts of personal data. The answer remains a lively social cons-
cience, buttressed by continued marketplace emphasis on decentralization of management and control over computer resources. There may be
need for the adoption of further, specific safeguards against surrepti-
tious entry into protected memories, but beyond that we should favor a
pluralistic and horizontal organization of the data marketplace.

Shared communal interest, in the sense of exposure to the World
Series or the Olympics or political party conventions, has been the
proud boast and genuine achievement of the broadcasting networks.
But the newer and more fragmented services, like cable and computer
conferencing, also cater to communities of interest—not all those who
live in Duluth, perhaps, but people who live in scattered places and
share an interest in astronomy. We are accustomed to using the telephone that way, and newspapers are increasingly deploying specialized
sections. To the extent that the continued viability of local newspapers
and local broadcasters is threatened by competitive services, the sense
of a shared local community with common concerns may be impaired.
It is too early to tell how serious may be the impairment, or what
alternatives may emerge as we become an information society, but it is
a subject of legitimate concern.

Equality of opportunity may suffer from the twin effects of competition and convergence. The elaborate system of internal cross-subsidies
developed by AT&T has been criticized as a form of unauthorized private government, but it did have the effect of extending telephone service on affordable terms to low-income and rural subscribers. That
may become a first casualty of the new competitive environment, not
for established services, but for some of the new ones. We may find a
communication surplus in the cities and a deficit in the country. To
avoid this, without reverting to monopoly organization of the markets,
could require the interposition of explicit government subsidies for particular services, such as health care and education, deemed particularly
valuable. The social-service satellite experiments sponsored by the Department of Commerce are an example of this trend. We are also going
to want to monitor the evolution of communication literacy for the
twenty-first century to avoid the emergence of deep and possibly resented social divisions.

Cross-cultural understanding should be promoted by the diversifi-
cation of media and the resulting push toward specialization of content.
The ability to hear jazz, blues, and "soul" music on the radio probably
contributed to a breakdown of racial segregation in the United States.
Still, there is a special concern arising from the "high priesthood" style

of computer communication. The report commissioned by France's Giscard d'Estaing, *The Computerization of Society*, said in effect that the person or country that controls the storage and processing of data has the capacity to impose a grammar and a dictionary on the rest of the world. This is perhaps a special instance of the concern over communication literacy, previously mentioned.

Freedom of expression could be endangered in the information society because of the progressive electronification of the media. As and when newspapers convert to distribution of their contents by electronic means—through computers to home video screens, let us say—the hands-off posture traditionally accorded print media could be replaced by the content regulation applied to broadcasting. This does not have to happen; the reverse could be the case, and broadcasting could be freed from content controls. But the evolution bears watching, for it could inject blandness or caution into the news and could diminish publishing to a sanctuary for nothing but classics and pornography. On the other hand, there are some brighter prospects, principally that computer publishing and interactive cable could bring back the personal essay, available on demand.

Creative opportunity could be enhanced by the greater number and diversity of outlets. It has been reliably estimated that a videodisc will return its investment if it is purchased by between 10,000 and 20,000 people—very different numbers from the tens of millions of viewers deemed necessary for prime-time network broadcasts. It will become possible to appeal to specialized and minority tastes. We can think of the ease with which rock music groups now form, produce a record, disband, reassemble, and produce another record, to get an idea of the creative freedom that may soon become possible for video producers and performers. On the other hand, copyright protection becomes more difficult with xerography and cassettes and electronic libraries. And the cumulative influence of computers and calculators may be anticreative, with their invitation to digital logic shorn of imagery or irony or humor.

Responsiveness to consumer needs is one of the basic claims of a competitive environment, and justifiably so. But what happens to consumer satisfactions if one or another communication mode is destroyed or mutilated by the competition? No more newspapers, no more local broadcaster—or one unable to carry a full complement of its usual services? The public may be inured by the scare claims it has heard in the past, but the boy who cried "wolf" did eventually find one. There are no ground rules now for failing newspapers, except to allow them to merge. Does society have a stake? What are the distinctive social values of a newspaper beyond its ability to wrap fish? Should government be prepared to step in when it has already declared (vis-à-vis cable) that it will not do so for broadcasters? I suspect the answer, in the near term at least, will have to take the form of self-help through expansion into other communication fields. Look at Dow-Jones and the *New York Times* and Gannett, all of whom are using electronic distribution.

Social and service stability, finally, is a broader statement of the same concern. It has to do with the security of investment in telecommunications concerns, and with a reasonable security of employment. Partly we can say that a shift from monopoly to competition inevitably introduces some insecurity and that the growth of the telecommunications sector as a whole will benefit all of society. Partly too we can say that some shifts in employment are humane, for example, letting automated services take over repetitive tasks. Yet the *pace* of change is a proper social concern, for it makes a difference whether retirements go unfilled or a whole function is abruptly abolished. Much will depend, again, on whether a telecommunications firm is thinking entrepreneurially about its place in the sun and is planning changes in its activities sufficiently in advance.

What can we say, then, about government's role and responsibility in the information society? First, it is useful to classify the possible responses. One is economic regulation, which will continue to apply where natural monopolies in markets are not amenable to competition; local telephone service is an example. Another is prescriptive regulation for certain kinds of behavior deemed injurious to society, for example, invasion of privacy. But the really interesting classification, one to which we have been leading, is protective regulation to limit or prevent the convergence of telecommunications services.

Will government return to exclusionary regulation as a means of protecting competitors instead of promoting competition? This is not likely. There will be pressures in that direction—for example, the efforts by newspaper publishers to exclude AT&T from becoming a videotex information provider—and some of these may get through. But in the main, it is not necessary so long as communication enterprises seize the opportunity that is now theirs to act like vigorous and responsible entrepreneurs. And there are some definite costs to an information society if it slips back down the regulatory path we have tortuously ascended.

There *are* nonmarket values that need attending, and some of these are threatened by open competition. But regulatory abuses can be and have been committed in the name of "externalities" by those unwilling to accept the benefits of market forces. We have to remember that regulatory agencies suffer from human faults and that the facts available to them are all too often inadequate or out of date. To give such an agency the responsibility of balancing competition against social values would be to give it a task beyond its competence, and more important, to risk losing those benefits of innovation, efficiency, responsiveness, and rational pricing that go with competition.

Most if not all the adverse social effects of an information society can be dealt with by governmental interventions other than economic regulation or market exclusion. A quick catalogue would include the criminal law, copyright, banking, privacy protection, social-service experiments, and perhaps some carefully calibrated tax incentives. There are also antitrust laws to deal with structural impediments to competition. Perhaps there will be need, after we have gained sufficient experience, for new social legislation addressed to one or another specific apprehended harm. But any such proposal should be carefully monitored to assure sensitivity to marketplace effects.

For I am persuaded, in conclusion, that the surest and least damaging safeguards of our human values will come from marketplace competition itself. Think of these forces: market reputation, consumer resistance, and entrepreneurial inventiveness. As long as the consumer has a choice, as long as he or she can turn to another source for the service that is found wanting, and as long as you and I can raise our voices of social vigilance and have them heard—then the satisfaction of the human values we prize so dearly should be fairly assured.

Chapter Seven

The First Amendment Meets the Information Society

William H. Read

I

Contemporary case law reflects concepts born of the Gutenberg revolution. Leading illustrations are two cases decided by the U.S. Supreme Court; one on broadcasting and the other on newspapers. A reexamination of those decisions—*Red Lion Broadcasting Co. v. FCC* (hereinafter cited as *Red Lion*) and *Miami Herald Publishing Co. v. Tornillo* (hereinafter cited as *Miami Herald*)—is offered here.[1] The purpose is twofold. The first is to argue that one of those cases would be decided differently in the light of the emergence of information societies. The second is to suggest that, should the Court revisit this area, any new holding ought to be conditioned by the fact that the Gutenberg revolution that created the foundation for the "press clause" of the First Amendment is giving way to the information society that is blurring boundaries between broadcasting and the press, on

For permission to reprint this article, I would like to thank Anthony Oettinger of Harvard's Program on Information Resources Policy and the President and Fellows of Harvard College.

the one hand, and "institutional media" and other modern commu-
nicators, on the other.

II

The information society is not likely to compel a wholesale reappraisal
of First Amendment law. True, the advent of new communication
technologies such as cable television, Xerox copiers, and teleprocessing
have had a sweeping impact on the laws of copyright and privacy.[2]
And other applications of electronic technology[3] may force change
elsewhere, too. Electronic eavesdropping by foreign agents may, for instance, become intolerable and may give rise to the passage of protective statutes or negotiated international conventions.[4] But just because
there will be more abundant and versatile means of communication in
the information society does not necessarily mean that the courts will
find it necessary to fundamentally rethink all the rules.

The information era does put some once-settled questions at issue
again. Among them is the judicial notion that a greater scope of government regulation is permissible with broadcasting than with the
press.[5] That notion is embedded in a two-track legal approach to mass
media: one track for publishers, another for broadcasters.

For print, the First Amendment stands behind the idea that no government, federal or local,[6] has the authority to interfere with the right
of a publisher to print whatever information that person cares to put
on paper. This is not to say that eighteenth-century images of a per-
secuted John Peter Zenger fighting to protect his press inspired courts to
insist on anything like a complete arm's length relationship between
press and government. Government can, and does, regulate publishers
in much the same manner as it controls other business enterprises.[7]
But even though antitrust, labor relations, and other business-oriented
laws present some limitations, reporters, writers, and editors carry out
their publishing functions generally free from repressive or chilling
governmental controls.[8] This is so because the press has come to
assume a special place in this country. Its role is popularly thought of
as the independent Fourth Estate, watchdog of the other three and
profit-making servant of an informed electorate.

While the basic law sanctioning freedom of the press is the two-
centuries-old First Amendment, judicial interpretations of that constitutional provision are of more recent vintage. Only after the outbreak of
World War I and the radical agitation that followed it did the U.S.
Supreme Court begin to search for coherent theories to explain either

tolerance or suppression of expression in specific cases. The early judicial theorists of note were Justices Holmes and Brandeis, who, in the absence of a "clear and present danger,"[9] favored "free trade in ideas."[10] The benefit of such a marketplace was later stated by Judge Learned Hand in these words: "[Right] conclusions are more likely to be gathered out of a multitude of tongues, than through any kind of authoritative selection."[11]

That government must keep its hands out of the editorial process of print media was made unequivocally clear a half century after the Supreme Court first began to explore the contours of the First Amendment. At issue in *Miami Herald*[12] was a Florida right-of-reply statute, which provided that if a candidate for office was attacked in a newspaper, the newspaper must offer rebuttal space to the offended candidate. A unanimous Supreme Court struck down the Florida statute in 1974, saying in part:

> The choice of material to go into a newspaper, and the decisions made as to limitations on the size of the paper, and content, and treatment of public issues and public officials—whether fair or unfair—constitutes the exercise of editorial control and judgment. It has yet to be demonstrated how governmental regulation of this crucial process can be exercised consistent with First Amendment guarantees of a free press as they have evolved to this time.[13]

This thumping endorsement for a free print press stands in sharp contrast to an earlier but still recent view of the Court with regard to broadcasting. The judicial position there upholds the right of government to interfere in programming decisions of radio and TV station operators. It is a position grounded in a tradition apart from that of print media. That tradition began in 1927 when it was established by statute that the electronic media (then only radio) have no right to exist without a federal license.[14]

The licensing requirement runs counter to a basic tenet of the First Amendment[15] because it mandates prior official approval before anyone can communicate by radio or television. Licenses are issued only to those persons who federal regulators determine will best serve the "public interest."[16] Moreover, the licensing scheme sets into motion controls and influences on the content of broadcast programs that surely would be held unconstitutional if applied to newspapers, magazines, and books. The most striking of these is the so-called Fairness Doctrine[17] that commands broadcasters to put balance in their public affairs programming and to provide air time to persons or points of view that the licensee might otherwise ignore. The Fairness Doctrine was upheld by the U.S. Supreme Court in 1969 in the *Red Lion* case. The Court offered this rationale: "It is the purpose of the First Amendment to preserve an uninhibited marketplace of ideas in which truth will ultimately prevail, rather than to countenance monopolization of that market, whether it be by the Government itself or a private licensee."[18] The premise upon which this statement rests was the Court's acceptance of the argument that broadcast media are fundamentally different from print media. Thus the Court reasoned: "Differences in the characteristics of news media justify differences in the First Amendment standards applied to them."[19]

What makes broadcast media different from print for constitutional purposes is the reliance of the former on the radio spectrum, or airwaves, for carrying information from senders to receivers. The spectrum is a natural and limited resource, something like navigable waterways, which generally are not considered to be suitable for private exploitation. When radio first took hold in this country, there was chaotic competition to exploit this finite resource until the federal government effectively stepped in to control the technology and allocate frequencies.

More than technical controls soon were to evolve, although these too were premised on the condition of spectrum scarcity. Because only a few could be licensed to broadcast, the idea developed that broadcasters were public trustees and therefore subject to governmental regulations.[20] That idea came to be bolstered on two counts: profits and power. As it turned out, the spectrum was a valuable resource that government had "given away." So it seemed proper to many that broadcasters should "pay back" in kind, ergo with programming not merely of commercial value but of public interest value. Furthermore, the supposed but never quite proven influence of broadcast media, especially television, over political and cultural attitudes was perceived by some as being too potent to be left wholly in the hands of media merchants. Thus by statute and regulation, held to be constitutional by federal courts, the legal scheme for broadcast media fits a different mold from that of print media.

From an intellectual, not to mention practical, standpoint, the two-track legal system for mass media seems unsatisfactory. In effect we have an eighteenth-century standard for one medium and a twentieth-century standard for another. While many commentators favor the older view for both, a substantial argument is on the record for unifying the law of mass communication under something like the Fairness Doctrine approach.[21] For proponents of either view, however, the Supreme Court erected a high barrier in *Red Lion* with its "differences in characteristics . . . differences in . . . standards" approach.

Are the media, print and broadcasting, truly different? As yet the only difference of constitutional dimension is that one medium uses the finite natural resource known as the radio spectrum. But, as has been mentioned, economic and social reasons also are said to support a two-track approach. The validity of all these reasons has been questioned in the past. More challenges can be anticipated in the future. For the fact is that as the information society further unfolds, the divergent legal approach to mass media makes less and less sense. To explain why, we can look at each of the above reasons in light of existing realities and trends.

II

Social Perspective

The argument for government regulation of media from a *social viewpoint* is the belief that the media in general, and television in particular, exert a powerful infuence over cultural and political attitudes. Some jurists have seen this as an acute problem. In the decision upholding the power of the Federal Communications Commission to regulate broadcast commercials for cigarettes, the U.S. Court of Appeals for the District of Columbia said:

> Written messages are not communicated unless they are read, and reading requires an affirmative act. Broadcast messages, in contrast, are "in the air." In an age of omnipresent radio, there scarcely breathes a citizen who does not know some part of a leading cigarette jingle by heart. Similarly, an ordinary habitual television watcher can avoid these commercials only by frequently leaving the room, changing the channel, or doing some other such affirmative act. It is difficult to calculate the subliminal impact of this persuasive propaganda, which may be heard even if not listened to, but it may reasonably be thought greater than the impact of the written word.[22]

Since the D.C. Appeals Court offered these comments in 1968, all cigarette advertising on radio and TV has been banned. Since then, however, the consumption of cigarettes has continued to pose a serious health hazard to the country, according to the U.S. Secretary of Health, Education, and Welfare.[23] Does this suggest that print media, which subsequently carried nearly all cigarette advertising, are more powerful than the appellate court had thought? Or is there possibly another explanation? Could it be that smokers enjoy their cigarettes and light up for pleasure and not merely in reponse to "pervasive propaganda"? Countless millions of Chinese do that every day, even though they never have seen a cigarette ad, either in print or on television.

The evidence is equally ambiguous with respect to the impact of broadcasting on politics. Not long ago, the idea spread that media were responsible for the expanded powers of the President.[24] Not only was it contended that media helped create an "imperial Presidency" but that television caused such a shift to the national level of American politics that only Washington politicians could hope to reach the White House.[25] That supposed "trend" was broken by Jimmy Carter, who upon reaching the White House found that the pendulum of power had swung back toward the Congress—"presidential television" notwithstanding.

A proper, indeed sensible, answer to those who advance the notion that media exert so much influence on the public mind that government must control them came from Justice Douglas:

> To say that the media have great decisionmaking powers without defined legal responsibilities or any formal duties of public accountability is both to overestimate their power and to put forth a meaningless formula for reform. How shall we make the *New York Times* "accountable" for its anti-Vietnam policy? Require it to print letters to the editor in support of the war? If the situation is as grave as stated, the remedy is fantastically inadequate. But the situation is not that grave. The *New York Times*, the *Chicago Tribune*, NBC, ABC, and CBS play a role in policy formation, but clearly they were not alone responsible, for example, for Johnson's decision not to run for re-election, Nixon's refusal to withdraw the troops from Vietnam, the rejection of the two-billion dollar New York bond issue, the defeat of Carswell and Haynesworth, or the Supreme Court's segregation, reapportionment and prayer decisions. The implications that the people of this country— except the proponents of the theory—are mere unthinking automatons manipulated by the media, without interests, conflicts, or prejudices is an assumption which I find quite maddening. The development of constitutional doctrine should not be based on such hysterical overestimation of media power and underestimation of the good sense of the American public.[26]

Economic Perspective

In the economic perspective too, the evidence is at least ambiguous on the question whether substantial and significant differences exist between print and broadcast media. Supporters of broadcast regulation make much of government's role in saying who will operate a station. While this is true, it is also the case that newspapers do not exist solely as a result of free market decision-making. As Justice Stewart has pointed out, "newspapers get Government mail subsidies and a limited antitrust immunity." Because of this, Justice Stewart has written that "it would require no great ingenuity" to make the same argument for newspapers as has been made for broadcasting.[27]

Nor does it take much ingenuity to stand the scarcity argument on its head. While the Court in *Red Lion* relied on spectrum scarcity to distinguish between media, the crucial reality for both print and broadcast media is "economic scarcity." A clear indicator of this is the trend toward the one newspaper town.[28] In broadcasting, too, there are constraints on the number of economically viable outlets. In an indirect but underemphasized manner, this was acknowledged in *Red Lion*:

> We need not deal with the argument that even if there is no longer a technological scarcity of frequencies limiting the number of broadcasters, there nevertheless is an economic scarcity in the sense that the Commission could or does limit entry to the broadcasting market on economic grounds and license no more stations than the market will support.[29]

Another economic argument is based on the fact that entry to the broadcast market is denied to some. It is said, therefore, that the Fairness Doctrine is essential to satisfy the claims of those excluded from operating stations. This argument finds some support because there is, after all, an absolute time limit on the amount of material that can be broadcast. In a practical sense a similar sort of limitation exists for newspapers. And this, too, has been acknowledged by the Supreme Court.

> It is correct ... that a newspaper is not subject to the finite technological limitations of time that confront a broadcaster but it is not correct to say that, as an economic reality, a newspaper can proceed to infinite expansion of its column space to accommodate the replies that a government agency determines or a statute commands the readers should have available.[30]

So when the economics of publishing and those of broadcasting are examined, what emerges is support for a conclusion reached by Justice Douglas in *Columbia Broadcasting System, Inc. v. Democratic National Committee* (hereinafter cited as *Columbia Broadcasting*).

> ... the press in a realistic sense is likewise not available to all ... the daily papers now established are unique in the sense that it would be virtually impossible for a competitor to enter the field due to financial exigencies of this era. The result is that in practical terms the newspapers and magazines, like TV and radio, are available to only a select few.

But what about the idea that, absent government control, greedy broadcast executives would commercially exploit the medium, with minimal regard for public service? The trouble here is that newspaper executives are, in large measure, of the same breed. In so-called cross-ownership situations,[32] top managers both publish newpapers and operate broadcast stations. And even when this is not the case, commercial motivations apparently can be strong in either publishing or broadcasting. Again, Justice Douglas provides a useful insight:

> TV and radio broadcasters have mined millions by selling merchandise, not in selling ideas across the broad spectrum of the First Amendment. But some newspapers have done precisely the same, loading their pages with advertisements: they publish, not discussions of critical issues confronting our society, but stories about murders, scandal, and slanderous matter touching the lives of public servants . . . .[33]

Finally, and most important, comes the question of monopoly. The Supreme Court in *Red Lion*, it may be recalled, spoke of a First Amendment that promotes "an uninhibited marketplace of ideas . . . rather than countenance monopolization of that market . . . [by] a private licensee."[34] As a practical matter there are, of course, far more broadcast stations on the air each day in this country than there are daily newspapers; three times as many, in fact.[35] But a more telling point comes from Bruce Owen, an economist who has analyzed media industries.

In examining the changing structure of the newspaper industry—an industry in which the number of daily and weekly newspapers has declined steadily since 1900, while the number of one-newspaper towns has risen—Owen concluded that restructuring came about in part because of "the introduction of new competing advertising and consumption technologies—motion pictures, radio, and television."[36] In other words, from an economic perspective, the mass media *compete* with each other. Indeed, according to a study produced by the Harvard Business School, "competition [for newspapers] from the electronic media had become particularly strong in large metropolitan markets."[37] And this is not just for advertising dollars. The Harvard study reported:

> Some industry analysts believe that VHF-TV stations with their network news programs had severely affected afternoon daily papers in some markets. Newspapermen claimed that TV could not provide in-depth coverage of the news, particularly local news. A one-hour television news show presented the equivalent of only a page or two from a newspaper. In the final analysis, the various media were competing for the time the consumer had to receive the information that she/he desired.[38]

If print and broadcasting compete with each other, as Owen says that they do, and if that competition is essentially for the time of "information consumers," as the Harvard Business School study finds, then, on economic grounds at least, the legal line drawn between media is weak, perhaps even arbitrary.

Technological Perspective

At the heart of the distinction drawn by the U.S. Supreme Court between the press and broadcasting is the belief that "the broadcast media pose unique and special problems not present in the traditional free speech case." What the Court finds to be "special" is the technological nature of the medium. Because broadcast frequencies are finite (and thus for reasons of efficient use must be allocated), the Court concluded in *Red Lion* that "it is idle to posit an unabridgeable First Amendment right to broadcast comparable to the right of every individual to speak, write or publish."[39]

The advent of cable television, over which the FCC has assumed jurisdiction[40] and has regulated in a mold similar to that as over-the-air TV, gives rise to an obvious counterargument. "Economic scarcity," as opposed to "physical scarcity," is the inherent condition of cable. In other words, the spread of cable depends, absent government regulations, entirely on market forces and not on an orderly allocation of the spectrum. Cable is more like telephone technology in this regard, although the concept of regulating cable like a common carrier has been rejected. Nonetheless, as cable gains greater acceptance around the country it adds to growing pressures for a unitary legal approach to mass media.[41]

Of greater significance in the information society is the convergence of broadcast and print technologies. Yes, convergence. Owen has a word for it: The print media are experiencing "electronification."[42] Electronic technology, Owen points out, is gradually replacing mechanical technology in printing. "In the past 10 to 15 years," he writes, "the electronic revolution has finally begun to be applied to the technical process by which newspaper and magazine copy is printed."[43] Reporters and editors, for example, are writing and processing news stories on televisionlike devices called video display terminals (VDT). And electronic transmission of even color pictures to distant printing plants has arrived. The *Wall Street Journal* and *Time* magazine are using the same basic technology to produce and nationally distribute their products as do the television networks. The technology includes the use of satellites, which, like over-the-air broadcasting, requires spectrum allocation and is therefore regulated by the FCC.

The convergence of print and broadcasting is perhaps best revealed by the advent of "teletext." Teletext systems transmit computerized information that is electronically displayed on a television screen. That these systems technologically transcend media is evidenced by the fact that they are operated by all sorts of communication organizations. In Britain, for instance, there are three systems: one run by the British Post Office, another by the BBC, and a third by a consortium. Similar services are being tried in six other countries, including three systems here in the United States—two are broadcast types being tested by the Public Broadcasting Service, while the other is operated by Reuters News Agency over the Manhattan Cable TV Network. Teletext is a leading example of what has been called the "still poorly charted area of 'media convergence,' where newspaper publishers, broadcasters and Post Offices and PTTs [government-operated telecommunications companies] find their interests overlapping and possibly conflicting."[44]

The convergence of print and broadcast technologies actually is part of a larger pattern. Anthony G. Oettinger has shown that the basic convergence is between communication technology on the one hand and computer technology on the other. That convergence, which Oettinger has labelled "compunications," has blurred the boundaries between the once distinct telecommunications and computer industries.[45] This in turn has caused a major regulatory headache because by law telecommunications has developed as a regulated monopoly, whereas the computer industry grew up in an intensely competitive marketplace.[46]

The impact of technological convergence makes for a rather interesting comparison. From a regulatory viewpoint, the computer and publishing industries have developed largely outside the ambit of direct government control. The contrary has been the case for telecommunications and broadcasting; both have been highly regulated, indeed by the same agency, the Federal Communications Commission. With convergence, the same basic technology is common to all these industries. The FCC, however, has statutory responsibility for two of the industries, not all four. Broadcasting and telecommunications are within the agency's jurisdiction; computers and print are outside.

This is not to suggest that all should be within, or all should be outside, the FCC's jurisdiction. From a technological perspective, however, it seems clear that these industries are bound to become further entangled. Evidence of this goes beyond teletext. At a meeting of the American Newspaper Publishers Association, AT&T and IBM presented their versions of the technological outlook for the years ahead. Essentially publishers were told that digitizing information—

putting pictures, data, news, graphics, and so on into computer form—and then communicating that information at lighting speeds within manageable costs are the two principal trends of great significance for everyone interested in the communication of information.

What do those trends mean? An answer came from Robert G. Marbut, chief executive officer of a major newspaper and broadcast chain:

> Regardless of the medium—television, CATV, newspapers, magazines—the technology to gather and process information and get it to a certain point along the distribution channel will be essentially the same for all... Common equipment will be used by all media. Whether it be VDT's [video display terminals], computers or broad band communications, we can expect all information providers to take advantage of this new technology. What I'm saying is this: The same technology is available to all information providers.[47]

The upshot of all this is clear: The technological underpinnings of *Red Lion* are eroding as the technological distinctions among media blur. To cling to spectrum scarcity as a rationale for a divergent legal approach to broadcasting is no longer viable. Indeed it is risky. For as newspapers and magazines more and more come to rely on satellites and other regulated communications technologies the danger exists that they will be drawn into the regulatory web. An argument can be made that this would be in the national interest. But if the national interest is otherwise than "in order to avoid regulating the content of print media," says Owen, "we may have to start now to deregulate the electronic media."[48]

III

The conclusion to be drawn from the foregoing is simply this: The rationale for a divergent, two-track legal approach for mass media has eroded. Once seemingly clear distinctions between print and broadcasting are no longer clear; "blur" is fast becoming an appropriate word. The question then is whether, in an information society, both media should be placed under the print standard or under the broadcast standard? Or, perhaps, a standard yet to be developed?

The progression of case law would suggest the print standard. *Red Lion* fixed the broadcast standard in 1969. Four years later, the Supreme Court seemed to be having second thoughts, or at least a multiplicity of thoughts, for in *Columbia Broadcasting* the Court gave us more of a symposium than a clear decision. In over one hundred pages of opinions, the justices revealed just how divisive the First Amendment can be. The case is notable for its range of views, not for any clear and concise judicial rule making. The issue in *Columbia Broadcasting* was whether a broadcast licensee could refuse to sell air time for political messages. The FCC said the licensee could. A divided and opinionated Supreme Court agreed.[49] Then, in its next term, a unanimous Court in *Miami Herald* held that government may not intrude into the function of editors in choosing what materials go into a newspaper and in deciding on the size and content of the paper and the treatment of public issues and officials.[50]

Curiously, the Court in *Miami Herald* never cited the *Red Lion* decision. Nor did the Court give any indication of the reason for this omission. The Court did present a clear interpretation of the press clause of the First Amendment on the issue of right of reply. Since, as has been shown, press and broadcasting are converging, it seems logical to say that the *Miami Herald* rule ought to be extended to broadcast media.[51] That conclusion was reached earlier, on other grounds, by Justice Douglas in *Columbia Broadcasting*:

> ... TV and radio stand in the same protected position under the First Amendment as do newspapers and magazines. The philosophy of the First Amendment requires that result, for the fear that Madison and Jefferson had of government intrusion is perhaps even more relevant to TV and radio than it is to newspapers and other like publications.... What kind of First Amendment would best serve our needs as we approach the 21st century may be an open question. But the old-fashioned First Amendment that we have is the Court's only guideline....[52]

IV

Aside from concluding that a single standard should be applied to both print and broadcasting, Justice Douglas also put his finger on an important question, one that leads to a postscript in this constitutional inquiry. What kind of First Amendment will this country need in the information society? That is an uncomfortable question because the performance of the Supreme Court so far has been less than satisfying in this area of law. As yet the Court has not provided a solid sense of just what the "free press" clause means.

The problem of the past is that the Court never attempted to confront the communication process as a whole. At times its analysis focused on who was communicating, as in cases dealing with the privilege of newsreporters,[53] at times the focus was on content, as in obscenity cases,[54] and at times the focus was on the individual's right to know, as in commercial speech cases.[55] By shifting the focus, even identical issues before the Court could have different outcomes. Both *Red Lion* and *Miami Herald* were, after all, right-of-reply cases. No right attached in *Miami Herald* under an analysis that emphasized freedom to communicate; a right did attach in *Red Lion* where the emphasis was placed on the public's right to know. The once seemingly viable "different standards for different media" reasoning facilitated contradictory results. With that reasoning now being technologically undermined, it seems prudent to recall that the communication process is exactly that: a process. For there to be completed or actual communication, as Claude E. Shannon demonstrated in his classic diagram of the process,[56] there must be a source of information, transmission, and reception. Absent any one factor, there is no communication.

It may be argued that the Constitution does not reach all these elements. Conceivably the "press clause" can be read as meaning that publishers have a right to print whatever they like and that right exists in a vacuum. Yet this is hardly realistic. Even Justice Stewart, who takes the position that "the publishing business is . . . the only organized private business that is given explicit constitutional protection."[57] recognizes by implication that publishers do not exist in a void. What they print is meaningless unless read. Otherwise, for Stewart, the guarantee would not serve "to create a fourth institution outside the Government as an additional check on the three official branches."[58]

Stewart's view raises a point that deserves some consideration. A democracy that worries about abuse of power certainly is well served by an additional check. But the notion of a Fourth Estate is not altogether trouble free. For one thing, it is becoming more and more difficult to say who is entitled to constitutional protection under the press clause. From a legal viewpoint, a graduate of Columbia University's School of Journalism who goes to work for the *New York Times* certainly is a full member of the Fourth Estate. That person's classmate, who joins CBS, is almost certainly not. Behind this segregation is an outmoded belief that broadcasting and the press have embodied a different set of values. It used to be said that the line between broadcast news and broadcast entertainment was a very fine one. Today the same can be said about newspapers. Like broadcast stations before them, newspapers are realizing that their profitability, and in some cases their survival, depends on sound business judgment. As a consequence, editors no longer merely try to supply a diet of information they believe an informed electorate needs. They are more responsive to the fact that their business must be sensitive to the market. Even the venerable *New York Times* has created "Weekend," "Home," and other special sections, which some critics say are more like "all the news that's fit to sell" in the *Times'* tradition of publishing "all the news that's fit to print."[59]

For those who champion a Fourth Estate, "supermarketing" information is a troubling trend in American journalism.[60] It is a trend, however, consistent with realities that even the biggest communication company, AT&T, has found in the new era of communication. Ma Bell is striving to transform its "century-old business from a supply-oriented business to a demand-oriented one."[61]

What we are witnessing, then, is a new communication market. The information society offers a market where suppliers will no longer be able to afford to take consumers for granted. Whether its the marketplace of ideas that Holmes and Brandeis thought of, or the technological market that AT&T long has dominated, the communication marketplace is changing. The old-fashioned arrogant editor who informed the public as he saw fit today faces the same profound problem of readjustment as does the giant telephone monopoly that now must learn how to compete for customers.

Meanwhile, the thrust of applied technology is to put the ability to communicate into more and more hands. In an information society any person with access to a Xerox copier will have as much, maybe more, publishing power than did Benjamin Franklin and his fellow colonial printers for whose benefit the press clause was written. Any political candidate may, moreover, find a bank of telephones or a computerized mailing list to be a greater communication asset than the editorial endorsement of a newspaper.

In the information society, with its abundant and versatile communication, "institutional media" will no longer hold exclusive command over mass communication technologies. The era when printers became publishers and radio engineers were granted broadcast licenses is no more. The field is open as never before. This condition brings with it a need for fresh legal thinking. To illustrate, suppose the following:

> The legislature of State X has before it an omnibus tax reform bill that, if enacted, would give individual taxpayers some relief while greatly increasing the tax burden of corporations. Suppose further that the entire business community, including local media corporations, oppose the bill because of the detrimental impact it would have on their posttax profits.
>
> Do corporations have a First Amendment right to speak out against such a bill? Yes, said the Supreme Court in *First National Bank of Boston* v.

Bellotti.[62] But their rights differ. The corporation that owns the local newspaper can frequently and freely editorialize in that paper against the bill. The corporate broadcaster, however, cannot express himself over the air even once without triggering affirmative Fairness Doctrine obligations. But what about, say, the local telephone company? What if that corporation decides to put its views before the public by printing a political message on its customers' monthly bills, or more likely by inserting such a message in each billing envelope? As a matter of constitutional law, could the state's public utilities commission stop this practice? If not, could the state commission compel the phone company to distribute opposing views?

Under the decision in *First National Bank of Boston* a telephone company, or any other corporation, has First Amendment rights. But what is the scope of those rights? In the hypothetical situation posed above, is the corporate "mailer," like newspapers, free from right-of-reply requirements, or can phone companies, banks, and other businesses be fitted into the broadcast mold?

To my mind, such questions are not idle. The ability to communicate to masses of people is spreading beyond the "institutional media." Moreover, big business and big government are ever more in an adversary relationship with each other. Thus questions like those above are certain to come before the courts as the information era approaches. Indeed, Chief Justice Burger appeared to be provoking just such questions when he concluded his concurring opinion in *First National Bank* with these words: "In short, the First Amendment does not 'belong' to any definable category of persons or entities: it belongs to all who exercise its freedoms."[63]

Burger, it seems has done more than merely stake out a position counter to that of Stewart's. Instead of narrowly reading the press clause, as does Stewart, the Chief Justice proposes an expansive interpretation "because of the difficulty, and perhaps impossibility, of distinguishing, either as a matter of fact or constitutional law, media corporations from [other] corporations. . . ."[64] The point is well taken, for the media business has become big business and is today "a far cry from the fragile printing presses that the Bill of Rights was designed to safeguard."[65]

In the end we have a Supreme Court that has two strong voices for moving the First Amendment in opposite directions. To the question, What kind of First Amendment do we need? Stewart replies with an interpretation that would protect publishers alone, whereas Burger answers that the constitutional guarantee should go to all who communicate. The ultimate issue, then, reaches beyond whether we need a First Amendment that protects broadcasting as well as the press. The

larger question is whether each form of communication should be ju-
dicially considered in isolation or as part of a collection of competing
forms of communication.

That question is bound to be in the air as the First Amendment tries
to meet the telecommunications revolution. Surely they will meet.
Where and when is difficult to predict. It is the kind of situation that
brings to mind a comment by Holmes: "It cannot be helped, it is as it
should be." According to Holmes, "the law is behind the times."[66]

Notes

1. *Red Lion Broadcasting Co.* v. *FCC*, 395 U.S. 367 (1969); *Miami Herald
Publishing Co.* v. *Tornillo*, 418 U.S. 241 (1974).

2. A survey of privacy laws, including the Federal Privacy Act of 1974, has
been prepared by Seipp, *Privacy and Disclosure, Regulation of Information
Systems*, Publication P-75-8, Program on Information Resources Policy, Harvard University (1975). An overview of how the Copyright Act of 1976 relates
to mass media can be found in Zuckman and Gaynes, MASS COMMUNICATIONS
LAW, 262–83 (1977).

3. A distinction of importance is between pure and applied technology.
Technology by itself does not create a revolution in communication. While the
inventions of yesteryear and modern-day research and development projects
certainly have said what was possible, it is *applied technology* that says what is
feasible. And applications, in turn, have depended on many factors. Consider
the case of the first medium of mass communication, the printing press. Four
hundred years elapsed between Gutenberg's invention and the institutional-
ization of the mass newspaper, or "penny press" as it was initially called in the
United States. Why did it take so long to apply this new technology? Sociologist Daniel Lerner has offered these thoughts: "The answer is that the West was
transforming its social order completely in order to absorb the innovation of
printed information available to all who wanted it. Consider that this transformation involved at least three basic dimensions of any social process: (1)
literacy: enough people had to learn to read to make a penny press feasible; (2)
income: enough readers had to earn "disposable income" (a very modern concept) to make the penny press a paying proposition; (3) *motivation*: enough
readers with an extra penny had to want to spend it on information rather than
on cakes and ale or beer and skittles" (Learner, *Communications, Development, World Order*. Occasional Papers of the Edward R. Murrow Center of
Public Diplomacy, September 15, 1978, p. 4).

4. Senator Moynihan drafted a bill, *The Foreign Surveillance Prevention
Act*, aimed at curbing Soviet eavesdropping in the United States by electronic
means and under diplomatic immunity. Press release, Senator Moynihan, July
27, 1977.

5. That government could regulate broadcasting in ways that were forbid-

den with respect to the press is not a judicial but an excecutive and legislative innovation. For historical treatments of broadcasting and broadcast regulation see Barnouw, A TOWER IN BABEL (1966); and Coase, *The Federal Communications Commission*, 2 J. LAW & ECON. 1.

6. The First Amendment was applied to the states through the due process clause of the Fourteenth Amendment in *Gitlow* v. *New York*, 268 U.S. 652 (1925).

7. *Associated Press* v. *United States*, 326 U.S. 1 (1945) (holding that the antitrust laws can be applied against a news agency).

8. Depending on the issue, the U.S. Supreme Court has tended to follow either an absolute approach (see *Saia* v. *New York*, 334 U.S. 558) or a balancing approach (see *Konigsberg* v. *State Bar of Cal.*, 366 U.S. 36 [1961] in the First Amendment area.

9. *Schenck* v. *United States*, 249 U.S. 47 (1919).

10. *Abrams* v. *United States*, 250 U.S. 616 (1919) (Holmes and Brandeis dissenting). *Whitney* v. *California*, 274 U.S. 357 (1927) (Holmes and Brandeis concurring).

11. *United States* v. *Associated Press*, 52 F. Supp. 362, at 372 (S.D.N.Y. 1943) *aff'd*, note 14 *supra*.

12. *Miami Herald Publishing Co.* v. *Tornillo*, *supra*.

13. *Id.* at 258.

14. The Radio Act of 1927 created the Federal Radio Commission with power to issue broadcast licenses if the "public convenience, interest, or necessity will be served thereby" 44 Stat. 1162 (1927). The legislation followed a period in which then Secretary of Commerce Herbert Hoover unsuccessfully tried to bring some order to the allocation and use of frequencies. His efforts were thwarted in part by federal courts. *Hoover* v. *Intercity Radio Col*, 286 Fed. 1003 (D.C. Cir. 1923) (holding that the Commerce Secretary was without authority to withhold or renew a license). *United States* v. *Zenith Radio Corp.*, 12 F. 2d 614 (7th Cir. 1926) (denying the Secretary power to put restrictions on use of frequencies).

15. Historical roots of the "press clause" of the First Amendment are found in adverse reaction to the press licensing system that existed in England until Parliament refused to renew the Printing Act in 1695.

16. 48 Stat. 1064 (1934), as amended 47 U.S.C. 151 et seq. (1970).

17. The Fairness Doctrine evolved out of a series of decisions by the Federal Communications Commission and subsequently was endorsed by the U.S. Congress in its 1959 Amendments to Section 315(a) of the Communications Act (note 6, *supra*.) P. L. 86–274, 73 Stat. 557. The doctrine amplifies on the "public interest" standard by providing that broadcasters "afford reasonable opportunity for the discussion of conflicting views on issues of public importance."

18. *Red Lion*, *supra* at 390.

19. Id. at 386.

20. A trustee is not a true owner of property but has duties and obligations to administer property for the benefit of others. The "property" in question

here, the radio spectrum, is said to be an "inalienable possession of the people of the United State's and their Government." S. 2930, 65 CONG. REC. 5735 (1924).

21. Barron, *Access to the Press—A New First Amendment Right*, 80 HARV. L. REV. 1641 (1967). It is worth noting that although print and broadcasting are treated differently by the law in some respects, they do in fact share much in common. Private ownership is perhaps the most striking feature. Few countries in the world, even democracies, have placed broadcasting, as well as print, in the private sector.

22. *Banzhaf* v. *FCC*, 132 U.S. App. D.C. 14, at 32–33, 405 F. 2 ed 1082, at 1100–1101 (1968), cert. denied, 396 U.S. 842 (1969).

23. *New York Times*, January 12, 1979, p. Al.

24. *See, e.g.*, Cater, *Toward a Public Philosophy of Government-Media Relations*, ASPEN NOTEBOOK ON GOVERNMENT AND THE MEDIA 6 (1973).

25. Robinson, *American Political Legitimacy in an Era of Electronic Journalism*, TELEVISION AS A SOCIAL FORCE 97 (1975).

26. *Columbia Broadcasting System Inc.* v. *Democratic National Committee* 412 U.S. 94, at N. 3, 152 (Douglas concurring) (1973).

27. *Columbia Broadcasting*, *supra* at 94.

28. Owen, *The Role of Print in an Electronic Society*, COMMUNICATIONS FOR TOMORROW, 229, at 232 (1978).

29. *Red Lion*, *supra* at 416.

30. *Miami Herald*, *supra* at 256–57.

31. *Columbia Broadcasting*, *supra* at 159.

32. "Cross ownership" refers to common ownership of a newspaper(s) and broadcast station(s) in the same market. Some 150 media combinations exist in 44 states. *New York Times*, November 22, 1977, p. 23.

33. *Columbia Broadcasting*, *supra* at 161.

34. Note 25, *supra*.

35. *Columbia Broadcasting*, at 93.

36. Owen, *supra* at 233.

37. Harvard Business School, NOTE ON THE NEWSPAPER INDUSTRY, 4-376-082 (1975), at 4.

38. *Id.*

39. *Red Lion* at 388.

40. *United States* v. *Southwestern Cable Co.*, 392 U.S. 157 (1968).

41. Schmidt, *Programming and Regulation of Mass Media*, COMMUNICATIONS FOR TOMORROW, 191, at 213 (1978).

42. Owen, *supra* at 242.

43. *Id.* at 237.

44. Winsbury, *Newspapers' Tactics for Teletext*, INTERMEDIA, February 1978, vol 6, no. 1, at 10. See also in the same issue, *A Chart of Teletext Systems*, at 12, *How the Post Office Wants to Use Viewdata*, at 13, and *Teletext Arrives—But What is Teletext?* at 32.

45. Oettinger, Berman, Read, INFORMATION RESOURCES FOR THE '80s (1977), 1–146.

46. For an in-depth treatment of what has been called reformulation of tele-communications regulation, see Loeb, *The Communications Act Policy Toward Competition: A Failure to Communicate*, 30 DUKE L.J. 1 (March 1978).

47. Robert G. Marbut is President and Chief Executive Officer of Harte-Hanks Communications, Inc. The quotation is from Marbut, "The Future of Newspapers" (presentation to the 50th Production Management Conference of the American Newspaper Publishers Association Research Institute, St. Louis, June 8, 1978).

48. Owen, *supra* at 242.

49. Narrowly read, *Columbia Broadcasting* is consistent with *Red Lion* since the holdings in both cases affirmed decisions made by the Federal Communications Commission. For his part, Chief Justice Burger viewed the FCC's rule-making process as a dynamic one and cautioned his colleagues against freezing that process with a constitutional holding (*Columbia Broadcasting* at 132). In effect, Burger reinforced a line of reasoning taken in *Red Lion* where the Court put heavy emphasis on the fact that Congress had chosen to regulate broadcasting through a federal commission.

50. *Miami Herald* at 258.

51. Absent recognition of convergence, courts are likely to continue to per-petuate the two-track approach in the mistaken belief that meaningful dis-tinctions between the two media are still to be found. See Judge Tamm's com-ments in *National Broadcasting Company, Inc.* v. FCC, 516 F. 2d 1101, at 1193 (1974).

52. *Columbia Broadcasting* at 148 and 160.

53. See *Branzburg* v. *Hayes*, 408 U.S. 665 (1972).

54. See *Roth* v. *United States*, 354 U.S. 476 (1957).

55. See *Bigelow* v. *Virginia*, 421 U.S. 809 (1975).

56. Shannon's diagram is reproduced in Pierce, *Communication*, SCIENTIFIC AMERICAN, vol. 227, no. 3, September 1972, at 32.

57. A few months after *Miami Herald*, Justice Stewart spoke at Yale where he commented on the role of the press in American society. The excerpt is contained in Schmidt, FREEDOM OF THE PRESS VS. PUBLIC ACCESS (1976), at 238.

58. *Id.* at 239.

59. For a good discussion of why the *New York Times* has been altering its traditional product, see *The New New York Times*, NEWSWEEK, April 25, 1977, p. 84.

60. Bordewich, *Supermarketing the Newspaper*, COLUMBIA JOURNALISM REVIEW, September/October 1977, p. 30. While it can be argued that pro-fessionalism among journalists is a restraint on any trend away from the values inherent in a Fourth Estate, the fact is that journalism is hardly a well-developed profession. Indeed Irving Kristol has contended it is an under-developed profession. See Kristol, *The Underdeveloped Profession*, THE PUBLIC INTEREST, WINTER 1967, p. 36.

61. Speech by Charles L. Brown, then chairman-elect of AT&T, in TELECOMMUNICATIONS REPORTS, Vol. 44, No. 47, November 27, 1978, p. 8.

62. *First National Bank of Boston* v. *Bellotti*, 46 LW 4371 (1978).

63. *Id.* at 4381. In his concurring opinion, Chief Justice Burger began by saying: "I . . . write separately to raise some questions likely to arise in this area in the future."

The *Wall Street Journal*, for instance, wondered whether established newspapers would be "happier if Mobil, implementing an idea it has toyed with, goes beyond buying ads to buying whole newspapers?" See *Bellotti and Beyond*, WALL STREET JOURNAL, May 5, 1978.

Charles B. Sieb offered this thought: "Legal complexities aside, a free press is inextricably entwined with the freedom of each of us. To lose one is to lose the other. The trick is to keep sight of that central fact in the face of changing technology, changing economic structures and a society that is changing before our eyes." See *The First Amendment as Corporate Business*, WASHINGTON POST, May 26, 1978.

64. *Id.* at 4379.

65. A discussion of "the press as a business" is found in, Kampelman, *The Power of the Press: A Problem for Our Democracy*, POLICY REVIEW, Fall 1978, 7–40, at 10–13.

66. Holmes, COLLECTED LEGAL PAPERS (1921), p. 231.

Chapter Eight

Videotex and Teletext: The Problem of Regulation

Tony Rimmer

In the information society citizens called for jury duty will not go to a courtroom but will serve in their own homes by watching the trial on television. This piece of stargazing comes, not from Future Shock's radical wing, but from the Chief Justice of the Florida Supreme Court.[1] Chief Justice Arthur England suggests that the medium through which jurors would keep in touch with each other throughout their "jury room" deliberations is two-way television.

Musings about this kind of media future may be blue sky for the judiciary, but for the Fourth Estate, the electronic future is now. In-house systems for writing and typesetting are in use throughout the industry. That last fateful step, transmitting the publication from in-house directly to the subscriber's home instead of sending the paperboy out on his rounds, will be basic in an information society.

For the press, the change will be a dramatic step. News will continue to increase in cost. Perhaps on these grounds, the temptation to abandon the new technology may be appealing. But other features, more related to tradition than economics, caution against a change to electronic publishing.

From a First Amendment standpoint, the information society may ultimately reveal to us that when newsprint was tossed out, the baby went with the bathwater. The transition from a paper distribution system to an electronic one could move the press from a relatively regulation-free environment to one encumbered with broadcast and common-carrier oversight. This chapter is concerned with that oversight. Its objective is to review communication law as it applies to the electronic publisher and consider whether this form of publishing might bring the newspaper within the regulatory scope of the Federal Communications Commission (FCC).

Ironically, in electronic publishing there is a promise of media diversity that First Amendment theorists have long pined for and declining newspaper numbers have increasingly denied. But clouding that promise is a tradition of government intervention in the administration of the broadcast spectrum, something print media are not accustomed to and do not care to acquiese to. Yet once over that electronic barrier, a great variety of bonuses beckon, all promising either new profit opportunities or at least cost reduction in old ones. And a lot of that promise revolves around two-way television.

There is, argues one sociologist, a movement in the industrialized world toward individualization.[2] For the moment, the individual is surrounded by media geared to the mass; the forms and content of these media are oriented toward collective behavior patterns and not necessarily toward the individual. Yet publishers stand on the threshold of a development that could lead from mass communication to individual communication. It might be called "on-demand communication," and two-way television holds the key to it.

The rationale goes something like this. If a newspaper can be written and edited on a screen, why slow down the updating process by putting it on a printing press and then onto paper, when it could be transmitted directly to a TV set at home? If people in a newsroom can become accustomed to reading and writing on a TV screen, why shouldn't individuals in an information society comfortably acquire the same custom, particularly if the screen was paper thin, handheld, and portable?[3] And if that TV monitor had a two-way function, then a content supply unique to that particular subscriber could be programmed for a unique customer "profile."[4] Newspapers already do this with their "zoning" practices whereby their markets are divided into various demographic categories that might appeal to small advertisers who do not want to reach the newspaper's total circulation at the newspaper's highest advertising rate. Hence the suburban editions of a citywide newspaper. Specialized magazines do the same sort of thing, appealing to certain interest groups. But their specialization is only a smaller version of "mass," and it is one-way. There is no direct user response. The information age offers the target of the smallest mass, catering to an audience of one. And it has the potential to respond to the changing whims of that audience of one. But to reach that audience of one, the publisher must transmit over the air or through a cable or telephone line, and with each of these media the Communications Act[5] empowers the FCC to take a regulatory interest. Two examples may help illustrate what regulation could mean in an information society when publishers move from newsprint to the electronic medium. First, consider the following scenario:[6]

> The *Miami Telepress* is a teletext newspaper licensed to use a portion of the electromagnetic spectrum. It has run an editorial critical of Frank Lee Graft, a state representative who is up for reelection. Graft demands an opportunity to reply to the editorial. The newspaper editor reminds Graft that a Mr. Tornillo lost the same right-of-reply argument with the *Miami Herald*[7] back in 1974. He suggests that Graft try to buy an advertisement. Graft's lawyer calls later that day threatening to file a complaint with the FCC, since his client is being denied his rights under the FCC's Political Editorials Rule.[8] He reminds the *Telepress* editor that these rules were affirmed by the Supreme Court in *Red Lion* v. *FCC*[9] and that his client's argument with the news service is a broadcasting and not a newspaper matter, so *Miami Herald* v. *Tornillo* is not relevant. The editor mumbles some First Amendment codewords as he buzzes for the *Telepress* lawyer.

Next contemplate what Richard Nixon might have been able to do in the Watergate era with the *Washington Post* if it ran only as a teletext news service out of the *Post*-owned television station WTOP-TV. The former President's comment that "the *Post* is going to have damnable problems out of this one. They have a television station . . . and they are going to have to get it renewed,"[10] takes an even more ominous tone than it did when it applied just to a newspaper.

It has been observed that the information society means a transformation, a "merging" of print, broadcast and common carrier communication services.[11] The characteristics of these media are seen to look more and more alike. Indeed, it is suggested that the gathering and processing of information and getting it to a certain point down the channel of distribution will increasingly be the same, regardless of how the final product is delivered to the home.[12] CBS Chairman William S. Paley recently noted that distinctions between the two media are blurring and that the newspaper and broadcast industries now have a lot more in common than they might heretofore have realized.[13] Paley noted that the time had come when cooperation rather than competition in promoting mutual interests (i.e., the removal of government intrusion in the editorial process) should be undertaken by different media.

The FCC chairman at the time, Charles Ferris, addressed the same point when he asked what free speech rights might attach to "newspapers" delivered over telephone lines or TV signals.[14] Should they parallel those of common carriers with rights of access and entry, but without content review? Or should they be treated as broadcasters, with fairness obligations? Perhaps an even more radical idea for a chairman of a regulatory agency: Should they be governed by the absence of regulation, like print? Ferris saw the first task as deciding "what . . . that creature [is] that uses the telephone lines and television set to provide the type of information that traditionally is obtained from the newspapers?"[15] Given that Ferris, at the time of this speech, was chairman of the regulatory agency central to the interest of this chapter, his question will be accorded considerable weight here, and the answer to it will be the goal toward which this essay strives.

The Technology

A confusing array of acronyms make up the field, each referring to a particular teletext or videotex system in a particular country. Thus we have Prestel, Ceefax, and Oracle in England; Antiope and Titan in France; CIBS, CAPTAINS, VRS, CCIS, and Hi-Ovis in Japan, and Telidon and Vista in Canada.[16] In the United States we have Knight-Ridder Newspapers/AT&T's Viewtron in Coral Gables, Florida;[17] the Departments of Agriculture and Commerce's Green Thumb service to farmers in Kentucky;[18] Telecomputing Corporation of America's The Source out of McLean, Virginia;[19] Micro TV's Info-Text in Philadelphia;[20] the Columbus (Ohio) *Dispatch*'s CompuServe;[21] the OCLC/Banc One Channel 2000;[22] and the system with perhaps the highest profile of all, Warner Amex's Qube,[23] also in Columbus, Ohio, and slated for Cincinnati, Pittsburgh, and Houston. One worldwide count in 1979 noted forty such systems under trial.[24]

Some generic terms have bubbled to the surface of this frothy mix. Initially, they were viewdata and teletext and latterly videotex. Scholarly usage would suggest videotex and teletext as generic terms describing the two technologies this chapter is concerned with. The former provides a two-way capability and the latter, one-way. However, technical innovation in teletext is providing an interactive capacity which suggests that the collective term videotex might suffice for both technologies. This chapter refers to videotex and teletext as two separate technologies, to fall into line with industry usage in the United States. Traditionally, teletext has been thought of as a set of information frames transmitted in a regular cycle that the home TV set "grabs" out of the air and displays on the screen. Videotex has been seen as a system where the user can interactively identify and call up individual frames from a database without having to wait through a transmission cycle as in teletext. Thomas offers the following general definition for the two systems:

> Systems for the widespread dissemination of textual and graphic information by wholly electronic means for display on low cost terminals (often suitably equipped television receivers) under the selective control of the recipient, using control procedures easily understood by untrained users.[25]

Tyler has offered a four-category system that throws more light on potential applications of the technology.[26]

1. Narrow-band interactive teletext, often known as "wired teletext," using the telephone system or similar networks to distribute data under the user's control at relatively slow speeds. The British Post Office's viewdata system Prestel is of this type.
2. Broadcast teletext, in which textual and graphic information is inserted into the redundant intervals in broadcast television signals creating a stream of television frames of information from which the user can make his or her selection. Ceefax and Oracle in England and Info-Text in the United States are examples of this broadcast teletext. This is inherently a one-way service, where the user can select the information frame but does not have a general two-way communication capability at his disposal. Transmission is typically at the rate of around four frames per second.
3. Wideband broadcast or cabled teletext, employing the same principles as the broadcast teletext but achieving much greater capacity by allocating all (or much) of a complete video channel to transmitting the alphanumeric or graphic characters. The amount of spectrum bandwidth taken up by a color television signal on a cable system could accommodate up to 25,000 frames of information if a teletext system were substituted in its place.
4. Wideband two-way teletext, in principle, would be the ideal. Here transmission in both directions is achieved on a sophisticated switched cable television network. Qube is such a system.

Hybrids of these are not only possible but will clearly proliferate in information societies. For example, one hybrid might involve telephone transmission of the user's information requests, the response to which are then supplied by cable. Another hybrid could be a regularly cycled set of frames as in the one-way systems, but, in response to telephoned user requests, unique frames could be inserted into the continuing cycled set.[27] In a cable channel with greater frame transmission, speeds, and capacity, this hybrid could resemble a more complete interactive service. There is a wide range of possible applications of these systems.[28] They include utility meter reading and fire and burglary monitoring, opinion polling, interactive games, electronic mail and newspaper delivery, information retrieval services, business transactions, and classified advertising. As entrepreneurs who market these systems say, the possibilities are limited only by the imaginations of the people involved.

How these systems will get into the home and who will provide particular services on them has become a contentious issue in the United States as entrepreneurs scramble to protect their individual interests in a situation where it is feared that decisions being made now may fix the way this particular electronic technology develops from now on.

What does the user get with these new technologies? The way the two technologies, videotex and teletext, have developed in England is a useful model for distinguishing between the two. Ceefax is an over-the-air teletext system. A typical television signal does not use up all the bandwidth allocated to it on the electromagnetic spectrum, so extra signals can be inserted into the transmission. The teletext signal is inserted into what is referred to as the blanking interval. To receive the signal, a television set must have a teletext decoder either built in or added on. These decoders are estimated to cost from \$200 to \$300.[29] Decoder manufacturers claim that with mass production, this figure will come down. But there are some 30,000[30] teletext-equipped TV sets in use in England, and there are no indications that the decoder price is coming down.

When it is received, the signal can either occupy the entire screen or be inserted into the image of another program already on the screen. Thus closed-captioning services could be provided by teletext,[31] or news flashes could be seen as they are published yet not interrupt an ongoing program. Alternatively, a user can "search" what teletext has to offer by calling up a menu page with a handheld switching pad. From that menu, the user can select a page or frame number for material of interest. The teletext decoder, in effect, "grabs" the requested frame off-air from the cycle being transmitted, as it did with the menu page, and presents it on screen. A third choice is merely to read all the teletext frames at the pace at which they are transmitted, provided, of course, that the transmission is paced slowly enough.

Videotex, or Prestel, as the system is called in England, is an over-the-wire system, in this case by telephone. The decoder used in teletext is required to get the image onto the TV screen. A user calls up a menu frame using the same handheld pad as for teletext; he identifies the information desired and enters those frame numbers into the system, using the keys on the hand pad. In videotex, an interactive facility is available. The user is able to search for desired information through a database organized in a logical tree structure. The interactive facility is a limited one. There is, for example, no sophisticated character string search capability. Frames are presented on request. There is, however, more control than that of teletext where the user must wait through a transmission cycle for a desired frame.[32] Teletext, with its continuous transmission cycle, seems suited to brief, constantly updated information. Videotex is suited to less ephemeral material as well as updated information. Since frames can be supplied on demand, there is no limit to the size of the videotex database.[33] The picture, in terms of transmission media, is relatively clear in Britain; the picture in the United States is more clouded. Teletext is over-the-air in Britain, and Prestel is over-the-telephone wire. The British experience is clearer for three reasons. First, the British technology was developed in the early 1970s, and the range of services available reflects early design assumptions.[34] Second, in the case of Prestel particularly, a primary purpose seen for the medium was that of increasing the utilization of the telephone system.[35] Prestel is therefore confined to the telephone network in England, which is a narrow-band system with low speed/quality transmission rates. Third, the administrative agencies that developed the two technologies—the British Post Office for Prestel and the British Broadcasting Corporation for Ceefax,—are state and quasi-state agencies. The regulatory environment is perhaps more generous when the regulator is able to promote its own technological developments than might be the case in the United States where the regulator is more in the business of balancing competing commercial interests in the market place against some statutory standard.

In the United States, a third medium has access to the home alongside broadcasting and the telephone. That medium is cable.[36] It is not as pervasive in Britain. In the United States, cable complicates the videotex/teletext picture both from a technological and a regulatory point of view.

How will teletext and videotex get into the home in the United States in the information society? That question has been posed several times in the past. The answers tend to reflect either the times in which they were asked (when the technology was simpler) or where the British model was seen as transposing directly into the U.S. environment.

Thus there is little recognition of the evolution of the technology. The arbitrary way one particular medium, be it cable, telephone, or broadcasting, was recommended as the best medium for the United States seems to have taken little notice of the fact that test operations were already successfully establishing operations on other media. Grundfest and Baer,[37] for example, saw a problem with the frame-by-frame billing transaction cost actually exceeding the value of the information being marketed. Their suggested solution was to go with an established billing system, and they argued for a negotiated contract between an information service firm and a telephone carrier to use that carrier's billing system. Failing agreement, they suggested firms might petition the FCC or the states to set reasonable terms and rates for access to the billing system. All this is predicated on use of the telephone as the means of access into the home. But systems in the United States operate successfully with their own billing systems. The Source, although accessed by telephone, bills at a flat-time rate for service.[38] Qube, with its constant monitoring of customer cable service use (it scans all users every six seconds), would seem to have the software capability already in place for a frame-by-frame charge system. Zerbinos[39] speculated that cable would be the "safest place" to put teletext from a regulatory point of view because the FCC did not impose as much content regulation on cable as it did on broadcasting. But the Fairness Doctrine, for example, applies to cablecasting just as it does to broadcasting.[40] It could be argued that the common-carriage status of the telephone might be a "safer" transmission medium for videotex. Perhaps the blanking interval of a TV broacast transmission will come to be regarded as a common-carriage facility, and the broadcaster may be able to lease it to, say, a newspaper.[41]

In fact, just as it has been suggested that information service providers might petition the FCC to obtain access to a telephone carrier's billing system, so might an electronic publisher move to petition the FCC for access to a broadcaster's blanking interval. If ideas of spectrum scarcity[42] as a basis for regulation still survive, then the blanking interval could be a candidate for regulatory development if broadcasters do not move to develop it themselves.

In the final analysis, speculation about the "safest" place for electronic publishing may really just be academic indulgence. First Amendment considerations do not loom as large, at least with broadcasters, as do economic considerations.[43] The wide mix of telecommunications being experimented with as electronic publishing outlets, with little mention in the trade papers of their First Amendment "safety" aspects, would tend to confirm this observation.[44] Commercial interest will dictate the most appropriate medium for any particular publisher. First Amendment arguments will be called on later to sanctify that channel selection.

In defense of these earlier recommendations, the dynamic nature of telecommunications means that it was bound to become dated. The FCC's Second Computer Inquiry;[45] two court cases, *FCC* v. *Midwest Video*[46] and *NARUC* v. *FCC*;[47] and Communications Act rewrite attempts in the Congress[48] have either set new directions or reflect new constraints in the media environment in the United States. Variations on, and judicial appeals against, these developments in the future will just as surely date this present study's efforts.

An interesting sideline to this speculation about the medium of choice for electronic publishing is the maneuvering for "spheres of influence," which is apparent between cable and telephone interests. Temple[49] has noted the existence of the so-called Telco Agreement, arranged with the cooperation of the House Communications Subcommittee, between the National Cable Television Association (NCTA) and the U.S. Independent Telephone Association (USITA). In this agreement the two organizations agreed not to compete with each other in their respective traditional areas of business (i.e., plain old telephone service for USITA members and the retransmission of television programs for NCTA members). Subject to respecting each other's defined sphere of interest, the telephone carriers and the cable television systems reserved the right to provide any "telecommunications service"[50] in any location either by using their own distribution facilities or by leasing transmission facilities from one another. For "telecommunications service," read electronic publishing.

The National Cable Television Association President echoed this agreement when he spoke of an anticipated three levels of service. He noted plain old telephone service, which the telephone companies would provide, cable television service, which the NCTA members would provide, and a middle ground of information services (examples of which he gave as information retrieval, burglar alarms, meter reading, and teletext) where the two media would compete.[51] It should be noted that there is one large component of the telephone industry, American Telephone and Telegraph (AT&T) which is not a member of USITA and is not a party to the Telco Agreement. Measured across a variety of indices, USITA member companies make up only 18 percent of the U.S. telephone business.[52] AT&T makes up the rest. But if AT&T is not with the Telco Agreement in the flesh, it is at least there in spirit. Charles A. Brown, as chairman of AT&T, said as much in hearings before the House Subcommittee on Communications in 1979.

When asked by the subcommittee's counsel whether AT&T might object to a prohibition on its being allowed to retransmit over-the-air broadcast signals, Mr. Brown said that was the status quo, and AT&T was not concerned "as long as it did not prevent us from getting into other things in connection with home information service, retrieval of data . . . "[53]

AT&T's interest in home information and data retrieval services has sent all manner of interest groups to Washington praying for relief. The government's response to these supplications is a confused yet apparently quite radical redefining of the way communication regulation is being thought about in the nation's capital. And because there is this element of the radical about it, the issue is far from resolved. A look at some of these ideas for each of the three media,—broadcasting, telephone, and cable,—might be a useful approach in clarifying this.

Broadcasting

There is an irony with broadcasting. A general impression might be gleaned from trade papers that electronic publishing is going to break out in the United States on broadcast teletext. KSL-TV's service in Salt Lake City is in full-scale operation. Micro-TV's Philadelphia-based service is busy soliciting advertising support. There is the appearance of activity in the area. Yet, below the surface, there is little movement. Technical progress over standards has slowed to a snail's pace, with backers of the British and Canadian systems at loggerheads with CBS over that network's attempt at an end run with the French Antiope system it backs.[54] CBS has filed a petition with the FCC promoting a modified Antiope system as the industry standard. Responses have been filed from the supporters of the other systems.[55] The petition process will be an extended one, and a set of teletext standards may not be announced for some time.

If the question of content regulation is one that makes print journalists "shudder at the thought,"[56] then it is unlikely that newspapers will look en masse to teletext for their initiation into electronic publishing. Representative Lionel Van Deerlin, a promoter of several attempts to rewrite the Communications Act, has noted that since "broadcasters must comply with the Fairness Doctrine and equal time provisions for political candidates," teletext, since it is broadcast, would have to comply with those standards.[57] Given this constraint, it is likely that only broadcasters would find teletext appealing, since that is all they know. A review of the organizations testing teletext does suggest that the medium is confined primarily to broadcasters.

A further reason for the apparent lack of any movement on the regulatory front for broadcasting, and more particularly for teletext, can be found among broadcasters themselves. Attempts at rewrites of the Communications Act failed, to some extent, because of broadcaster opposition to the idea of a more competitive broadcast environment.[58] In fact, so effective was this broadcaster opposition that after H.R. 3333 failed, the next bill on the subject, H.R. 6121, contained only common-carriage provisions. Broadcasting matters were excluded from the bill, which concentrated instead on the presumably more popular deregulation of the telephone industry. Broadcasters apparently preferred the devil they knew, the Communications Act of 1934, to a deregulated broadcast system.

The Telephone System

The telephone system as the information society's medium for electronic publishing is interesting for two reasons: (1) because of the scale of the companies behind it; and (2) because of their interest in becoming publishers, as well as their traditional role as common carriers with facilities for lease.[59] Thus GTE has purchased U.S. rights to the British Prestel system;[60] AT&T is involved in the Coral Gable Viewtron test with Knight-Ridder[61] and has several tests of its own planned or in operation around the country.[62]

A clue to what is at stake is suggested by the possibility of AT&T's expanding its yellow pages into an electronic classified advertising service. This would be an unregulated data processing service with a lot of revenue potential. At stake is $4.6 billion, the amount that classified advertising made up of total newspaper advertising revenues in 1979.[63] If electronically distributed, classifieds could be updated continuously, and a user would need only call up those that were of personal interest. The vast memory banks available in computing today might also allow for more depth of information in each classified advertisement. Floor plans of advertised houses, for example, might be available in the realty section.[64].

Two tests in which AT&T is involved might seem to the consumer to offer similar services.[65] For example, both offer telephone directories and classified advertising from grocery and department stores. From a regulatory point of view, however, there are some critical differences. In the Coral Gables Viewtron experiment, AT&T is providing the telephone lines and terminal equipment for the users. Knight-Ridder is providing the computer, the software, and the database.[66] In the proposed trial in Austin, the telephone company plans to go further than the common-carrier function. It will provide the computer, database, telephone lines and terminal at the user's end, the so-called customer premises equipment (CPE).[67] In the Viewtron trial, the telephone company appears simply to be fulfilling its traditional role as a common carrier. But in the Austin trial, the common carrier is becoming an electronic publisher. In doing so, it seems to be flying in the face of both the FCC's latest regulatory orders in the Second Computer Inquiry and the philosophy behind regulatory thinking in Congress. In the process, AT&T seems to be confirming all the worst fears that the press has about its role as an electronic publisher in the information society.[68]

In its Second Computer Inquiry, the FCC radically redefined the way it looks at telephone service. The commission had previously tried to face up to the impact that computer technology and its market applications were having on communication common-carrier services and regulation. The phenomenon of data transmission and processing was playing havoc with old ideas based on voice transmission. In its First Computer Inquiry,[69] the FCC held hearings, in part, to try and clarify the situation. The outcome was a dual definitional scheme that distinguished between regulated communication services and unregulated data processing services. In the case of hybrids between the two, the definitional scheme was designed to consider the orientation of the service.[70] To prevent the possibility that common-carrier services might favor their own data processing activities when providing unregulated data processing services, a policy of "maximum separation" was initiated whereby a carrier had to furnish data processing services through a separate corporate entity.[71] That inquiry came from an era when central computers operated in conjunction with "unintelligent" terminals on customer premises. Technological development in data processing quickly outpaced the First Computer Inquiry. Advances in microprocessor and large-scale integrated circuitry have since allowed the construction of mini- and microcomputers. Customer terminals are now more intelligent. Distributed processing has become the norm; computers and terminals perform both data processing and communication control within a network and at a customer's premises.[72] In the Second Computer Inquiry, various new definitional schemes were discussed to try to specify more precisely what data processing was and what it was not, and how hybrids between the two might be accommodated.[73] The commission finally came down with a distinction between a *basic* transmission service and an *enhanced* transmission service. The former would be regulated in the traditional common-carrier sense, the latter unregulated. The basic service was limited to the offering of transmission capacity between two or more points suitable for a user's transmission needs and subject only to technical parameters of fidelity or distortion criteria. An enhanced service then became any offering that is more than basic service.[74] Typically, this might be where computer applications are used to act on the content or code of a subscriber's information. In an enhanced service, the "content of the information need not be changed and may simply involve subscriber interaction with stored information."[75] That is what a videotex service is. Thus, videotex is an enhanced service and therefore not regulated in terms of tariff regulation and having to get the FCC's prior approval for new services.

Is this good news for future electronic publishers in the information society? Perhaps. But the commission qualifies itself by reminding its audience that it has not given up jurisdiction over enhanced services, merely that it has cchosen not to initiate a comprehensive regulatory scheme for the service. Rather, it notes that if occasional problems concerning enhanced services arise that require the commission to invoke its subject-matter jurisdiction and intervene, the FCC would prefer to handle the resolution of those problems on an individual basis.[76]

Since carriers have traditionally "bundled" terminal equipment along with transmission facilities in providing a service, the commission next addressed the possibility of whether "bundling" could limit a user's freedom of choice in putting together a preferred service and equipment package. Trends in technology, the commission observed,[77] enable terminal equipment to function as an enhancement to basic common-carrier service, yet bundling has forced users to go with packages of transmission and equipment services that the carrier offers even though these may not meet the needs of the customer. Concluding that bundling inhibits competition, the commission moved in the Second Computer Inquiry to deregulate customer premises equipment. The analogy in present homes is that the rental charge built in for telephone receivers will now have to be itemized separately. This will, in effect, serve notice on customers that they may now purchase outright from other vendors the telephone equipment that meets their needs. To further complicate matters, the commission, using a standard that identified dominant carriers as those "telephone companies which have sufficient market power to engage in effective anti-competitive activity on a national scale and which possess sufficient resources to enter the competitive market through a separate subsidiary,"[78] ruled that AT&T would have to market terminal equipment and enhanced services through a subsidiary.[79]

For the Austin trial, the separate subsidiary rule suggests that AT&T (in this case Southwestern Bell) should be providing only the transmission facilities for the test. The computer, the customer premises equipment, and the database should be provided through a subsidiary or by another vendor.[80] Whether the FCC has the authority even to allow AT&T into the enhanced services market, contrary to the 1956 Consent Decree[81] that barred AT&T from offering services such as data processing and information retrieval, is hotly contested by the Department of Justice.[82] There is obviously still much to be resolved in the area.

Interest in the Consent Decree has also been manifest in the Congress where concern for the welfare of the Justice Department's antitrust activities against AT&T in effect sidelined the fortunes of H.R. 6121, a rewrite of the Communications Act.[83] The bill allows AT&T to compete in areas not regulated by the FCC, thereby modifying the 1956 Consent Decree. The bill also contains an American Newspaper Publishers Association (ANPA) sponsored amendment that prohibits "dominant carriers" (read, AT&T) from offering data retrieval services either by the parent company or its subsidiaries, for the types of information being provided by newspapers, periodicals, radio and television.[84] The scope of the amendement is wide. This amendment, too, suggests that the AT&T Austin trial is out of order.

But of course, like H.R. 3333, the bill that went before it, this is all still bill drafting rather than statutory law. How seriously congressional sentiment should be regarded on electronic publishing is not clear. The two bills failed for reasons other than mass media service amendments. Apart from the sweeping ban on the provision of mass media services, however, there is a lot that is similar between H.R. 6121 and the Second Computer Inquiry. Lionel Van Deerlin, the house bill's sponsor, conceded as much in his commendation of the FCC's work when the commission's inquiry report was released.[85]

Cable Television

It is in the context of cable television that most regulatory interest in two-way television has been seen. Unfortunately, the two-way facility has taken a back seat to arguments about the validity of the FCC's rules regarding access to cable.[86] In 1972 the FCC required new cable television systems to include two-way technical capacity.[87] That requirement (capacity for return communication on at least a nonvoice basis) was later modified to apply to new cable systems with more than 3,500 subscribers, and later still to older systems if they were rebuilding to comply with other FCC technical standards. Along with this two-way rule were others requiring mandatory access to cable facilities for the public, educators, and local government; the provision of studios and production equipment for the use of these access groups; and a system capacity of twenty channels. In *Midwest Video II*, these rules were struck down by the Supreme Court as exceeding the statutory authority allowed the FCC. The Court found that the rules required cable system operators to behave as common carriers.[88] If common-carrier obligations cannot be imposed on broadcasters, then the Court held they surely cannot be imposed on cable systems, arguing that the variant technology of cable did not lessen the journalistic discretion that Congress has consistently allowed broadcasters.[89] The Court did concede that less stringent access rules might get a more favorable hearing. But, despite an FCC plea that the access, capacity, and facilities rules be considered separately from one other, the Court concluded that the technical capacity rules were set up to promote the access rules and thus set them aside as a group, without considering the merits of each.[90] So the two-way rules failed without much of a hearing.

The two-way rules had not had much of a hearing from the FCC either. When the commission first promoted the two-way idea, it conceded that there may not be a demand for the service and that is why it never required more than the technical, rather than operational, capacity for two-way services. Ironically, it would now appear the two-way and access services are the sine qua non for obtaining cable TV franchises.[91]

One final appellate-level case has bearing on the plight of two-way television, and that is *NARUC II*.[92] It is included in this discussion for three reasons. First, it highlights a judicial finding that individual cable company services are severable, despite the FCC's tendency to consider all services together, and further that common-carriage regulation can be applied to any of those severed services.[93] Second, it highlights a predicament for the FCC about how far the agency can extend its interest into intrastate jurisdictions.[94] And third, it demonstrates that there is still a great deal of confusion about what two-way service actually is. The 1972 Cable Report called for capacity for return communication at least of a nonvoice quality. In terms of videotex, for example, this would refer to a keypunched request from a user for certain information to be displayed on the user's TV screen. It could also be read to imply that the signal from the computer to the user's screen is video. It does not clarify whether the two signals (i.e., to the computer and back to the user) are integrally tied as they presumably would be with videotex where the nonvideo return signal is calling up a video response from a head-end computer. In fact, whether a static image, which is all that videotex technology can offer at the moment, would constitute "video" is not clear. It may be a nonvideo signal. Signals that are not tied together integrally, or that are both nonvideo are not specifically accounted for.[95]

To further add to the confusion, one of the FCC's prerequisites for common-carrier status is that the system be such that customers transmit intelligence of their own choosing.[96] That is, the customer provides the information to be transmitted, the carrier provides only the transmission facilities. In videotex, only the return mode (from customer to computer) would be of the customer's choosing. The line from the computer back to the customer would consist of the computer's response to that intelligence. On the basis of the holding in *NARUC II*, the court would find videotex to be a common-carrier activity.[97] The other prerequisite to common-carrier status, holding out a service indifferently to all potential users,[98] would also be met by videotex.

A cable system offering a videotex service might, therefore, under the terms of *NARUC II*, be a common carrier, an appellation the FCC refuses to give cable systems. The court in *NARUC II* concluded "that most, if not all, of the uses to which the two-way non-video capability is likely to be put fall within the term "carrier" as used in 47 U.S.C. 152(b)."[99] Notwithstanding all of the foregoing, the court found that cable operators providing two-way intrastate nonvideo communication were acting outside the commission's jurisdiction because of the express provision of section 2(b),[100] which prohibits FCC involvement in intrastate matters. The Austin trial by AT&T would seem to meet that provision. It is two-way and intrastate.

Constitutional Aspects

The First Amendment rights of broadcasters and cable operators do not seem to loom large in the FCC's regulation of electronic publication. For example, using *Red Lion's* finding of a public speech right over that of broadcaster autonomy, the commission has enthusiastically sought to extend that public right of access to cable television at the expense of the cable operator's speech rights. The commission's brief[101] in *Midwest Video II* tersely dismissed the cable operator's First Amendment claims by saying that (1) cable systems retransmit broadcast signals (and are thereby "reasonably ancillary" to the FCC's interest in the regulation of broadcasting),[102] that (2) *Midwest Video I* authorized rules designed to achieve the commission's program diversity "objectives," and that (3) First Amendment goals are promoted by access rules, citing *Red Lion* language about an uninhibited marketplace of ideas and monopolization of that market. In order to impose access rules the FCC tried to use the retransmission of broadcast signals as a lever to gain authority over cablecasting and two-way services, even though these have little to do with over-the-air transmission services, which are the foundation of FCC broadcast authority.

Not everyone is quite so pessimistic about the government's intentions. From the affirmative First Amendment ideas of Emerson,[103] which achieved their ultimate expression in *Red Lion*, Silber[104] sees a trend that has developed over the years toward what he calls "broadcaster autonomy" in decisions such as *CBS* v. *DNC*,[105] where a broadcaster's refusal to carry an advertisement was upheld. Silber was bemused, though, by *FCC* v. *Pacifica*, through which the FCC used George Carlin's "seven dirty words" to sustain its proscription of indecency. The rationale in *Pacifica* for denying broadcasters the same First Amendment protection as for print was because broadcasting had "established a uniquely pervasive presence in the lives of all Americans,"[106] and that broadcasting is uniquely accessible to children, even those too young to read.[107]

This "captive audience" idea seems to be a variant of the "power of broadcasting" idea, which is usually listed along with the ideas of "public ownership of the airwaves" and broadcast frequencies being "scarce resources" as the three most common rationales for distinguishing broadcasting from print for regulatory purposes. And, where broadcasting goes, interest groups with memories extending back to *Red Lion* are trying to make sure cable goes.

With cable television, the idea that seems to get most of the attention is the scarcity theory. Public ownership as a basis for regulation works only for broadcast retransmission channels on cable through the "reasonably ancillary" doctrine. Access and two-way channels are more likely to be privately owned. The power idea, with its fear of the unknown, appears to be more theologically than empirically based, and lingers in the background rather than center stage for cable.[108]

Scarcity, the idea that a limited number of channels requires government assurances that a multiplicity of opinions will be telecast, should not, it has been argued, apply to cable TV because of its many channels. A structural or technological solution[109] to regulation is now considered possible compared to the behavioral, or content regulation, of the precable era. In fact, such is the optimism that diversity's utopia has now arrived that petitions have been filed by the National Telecommunications and Information Administration (NTIA) with the FCC to release cable systems from requirements under the Fairness Doctrine.[110] Needless to say, there are still pessimists who see in cable

television more of the same. Barron, for example, notes that technological solutions have so far not rendered his preferred social engineering
unnecessary, pointing out that so far most cable systems merely offer
more choices among the networks.[111]

A further problem for the scarcity theory as a basis for regulation is
that newspapers can be shown to be much more scarce than broadcast
and cable outlets, yet newspaper content is not regulated. In *Home
Box Office* v. *FCC*, the court, in reference to *Miami Herald* v. *Tornillo*, said that "scarcity which is the result solely of economic conditions
is apparently insufficient to justify even limited government intrusion
into the First Amendment rights of the conventional press...and
there is nothing in the record before us to suggest a constitutional distinction between cable television and newspapers on this point."[112]

The lack of sound theory supporting broadcast and cable content
regulation would suggest that it is time to put broadcast and cable
alongside the print model in First Amendment treatment. But tradition
does not lie down and die that easily.

Conclusion

Nirvana is not yet here for First Amendment theorists in broadcasting
and cable television. And it may be a long time coming. Despite the
carrots of deregulation, and movement toward the print model, which
the FCC chairman has offered broadcasters and cable operators in
his speeches in return for more competition,[113] some of the problems ahead suggest more rather than less regulation. Consider the
following.

1. Electronic publication may threaten broadcast news and thereby the FCC's
   policy of localism. Busterna, for example, has found that a ten point increase in cable penetration in a market is associated with a 7 percent de-
   cline in broadcast news expenditures.[114]
2. Euphoria over what appears to be the end of the scarcity theory as a basis
   for regulation takes little account of the fact that nearly all cable cities are
   one-company operations. Who the operator is may become an important
   determinant of who gets on the cable. If the Times-Mirror owns the cable
   company, what are the prospects for other newspapers of getting on that
   cable?
3. In the 105-channel operations being announced today, what might be the
   consequences of electronic newspapers being buried in a swamp of competing media? In the British Prestel system, for example, newspapers can
   be lost among a lot of other information services on just one channel. Is it
   conceivable that the ANPA may find itself in the ironic situation of appealing to the government for relief and arguing for another round of Failing

Newspaper legislation? H.R. 6121, the Communications Act rewrite attempt, with its sweeping restriction on who can electronically publish mass media services, while confining AT&T to directory services, may find it has applied a "Meiklejohnian" approach to defining mass media services (i.e., those "services designed to inform the electorate on issues of public concern.)"[115] If all else is left to dominant carriers, electronic newspapers may find that what the audience wants more of is not information "of public concern" but directory information, box scores, and the like.

4. Kaplan[116] has noted that what seem to be interviews (of "public concern"?) on Warner Amex's two-way Qube system turn out to be marketing exercises where the interviewees have actually bought time for the interviews. Kaplan asks whether this type of programming should be regulated, and if so, by whom?

5. The price of decoders is high enough that the possibility of taking advantage of the end-of-scarcity theory will be denied many people. In the electronic publication era, might the advent of this wonderful diversity of voices actually disenfranchise much of the audience?

6. An issue not considered in this study, that of privacy, becomes very relevant with the electronic surveillance necessary to keep track of a subscriber's media use,—for billing purposes, to rebuild user profiles, or for marketing purposes.

7. Questions concerning who the publisher is in libel cases and in the reuse of material have yet to be addressed.

In spite of the excitement surrounding trials and speculation in electronic publishing, price and privacy questions might make us all pine for the good old days when we had to retrieve a sodden but low-priced newspaper out of the tree in the front yard, and when the only privacy question of concern was whether we had remembered to put a bathrobe on before going out.

Notes

1. AUSTIN AMERICAN-STATESMAN, August 27, 1980, at F8.

2. Silberman, *Communication Systems and Future Behavior Patterns*, 24 INT. SOC. 337 (1977)

3. Rosenfield, *A Complete Electronic Newspaper?* 35 ANTIOCH REV. 171, 177 (1977). In-house electronic writing and editing systems for broadcast newsrooms are now coming onto the market. One such system, marketed by Station Business Systems, a subsidiary of Control Data Corporation, has been installed at WQAD-TV, Moline, Illinois, and at KSL-TV, Salt Lake City, Utah. Letter from George V. Pupala, director of sales, Station Business Systems (October 6, 1980).

4. Research Seminar by Wayne Danielson, "Micro-computers and Content Analysis," to Journalism Department, University of Texas at Austin

(October 15, 1980). Dr. Danielson noted how a user might put into a computer system a series of descriptors of personal subject interest to set up an individual news "profile." This profile might interrogate other data sets, such as those of an electronic publisher, for news and information of interest to that user.

5. Communications Act of 1934, 47 U.S.C. § 151 *et seq.*

6. Suggested by Zerbinos, "Teletext/Videotex and Freedom of the Press" (paper presented to the Association for Education in Journalism Annual Convention, Boston 1980.) (hereinafter Zerbinos), at 14.

7. *Miami Herald Pub. Co.* v. *Tornillo*, 418 U.S. 241 (1974).

8. 47 C.F.R. §§ 73, 1930 (1978).

9. *Red Lion Broadcasting Co.* v. *FCC*, 395 U.S. 367 (1969). But see Power, *Or of the (Broadcast) Press*, 55 TEXAS L. REV. (1976), for a commentary on the ups and downs of the fortunes of the consumer speech right enunciated in *Red Lion*.

10. Taped statement of Richard Nixon to H. R. Haldeman and John Dean (September 15, 1972), quoted in SENATE SELECT COMM. ON PRESIDENTIAL CAMPAIGN ACTIVITIES, FINAL REPORT, S. Rep. No. 981, 93rd Cong., 2d sess. 149 (1974).

11. Address by Charles Ferris, "New Technology and the Merging Media . . . A Time for Imagination" (annual meeting of Audit Bureau of Circulation, New Or-leans, La., November 7, 1979) (hereinafter Ferris, Merging Media). Ferris was chairman of the FCC at that time.

12. Watts, "Major Issues of the 1980's: First Amendment Implications of New Communications Technology" (presented to the Association for Education in Journalism Annual Convention, Boston, 1980) (hereinafter Watts). Watts is the staff counsel, American Newspaper Publishers Association (ANPA).

13. BROADCASTING, June 16, 1980, at 114.

14. Ferris, Merging Media, *supra.*

15. "Up to Speed at the Ferris FCC," interview with FCC Chairman Ferris, BROADCASTING, April 14, 1980, at 68.

16. Prestel and the other systems noted are described in Rimmer, "Viewdata—Interactive Television, with Particular Emphasis on the British Post Office's Prestel" (paper presented to the Association for Education in Journalism Annual Convention, Houston, 1979) (hereinafter Rimmer), at 5.

17. *Viewtron Test Started by Knight-Ridder*, EDITOR AND PUBLISHER, July 26, 1980, at 18; *AT&T/Knight-Ridder Teletext Trial Underway*, 12 E&ITV, August, 1980, at 13; Kelly, *All the New That's Fit to Compute*, WASHINGTON JOURNALISM REVIEW, April 1980, 13 (hereinafter *All the News That's Fit to Compute*), at 13.

18.Ayers, *The Greening of Kentucky*, TELEPHONY, April 14, 1980, at 28.

19. Ashe, *The Limitless Possibilities of Delivering Information Electronically*, ANPE BULLETIN, April 1980 (hereinafter Ashe), at 3; Sigel, *Videotext in the U.S.*, in VIDEOTEXT: THE COMING REVOLUTION IN HOME/OFFICE INFORMATION RETRIEVAL, Sigel (1980) (hereinafter Sigel), at 107.

20. Sigel, at 92, *supra* at 19.

21. *AP, Newspapers and Computer Firm Join in Test of Telephone Delivery*, BROADCASTING, July 7, 1980, at 49.

22. *OCLC Is Doing What?* LITA NEWSLETTER (Library and Information Technology Association), Summer 1980, at 5; *Bank to Try Out Two-Way Phone System in Columbus*, BROADCASTING, February 4, 1980, at 50.

23. Rimmer, *supra* note 16, at 8; *Special Report, Warner Cable's Qube: Exploring the Outer Reaches of Two-Way T.V.*, BROADCASTING, July 31, 1978, at 27.

24. Tyler, *New Media in the Information Economy: Prospects and Problems for Viewdata and Electronic Publishing* (hereinafter Tyler), at 266, in PROCEEDINGS OF THE SIXTH ANNUAL TELECOMMUNICATIONS POLICY RESEARCH CONFERENCE, ed. H. Dordick (1979) (hereinafter Dordick).

25. Thomas, "Current Developments and Trends in Videotex and Teletext" (paper presented to the IIC Annual Conference, London, 1979), at 10.

26. Tyler at 266, *supra* note 24.

27. Bonneville Broadcasting Corporation reports that it is developing a Touch-Tone Teletext that allows for a degree of interaction between the user and the teletext central computer. BROADCASTING, April 21, 1980, at 78.

28. Rimmer at 4, *supra* note 16, suggests some 70 possible uses.

29. BROADCASTING, August 4, 1980, 62, at 63, suggests a set-top decoder price of $200, less if built into the set; Tyler, *supra* note 24, at 270, prices the decoder at $300; NEW SCIENTIST, February 14, 1980, at 483, also prices it at $300.

30. BROADCASTING, December 17, 1979, at 38. Sales of decoder equipped sets were reported running at 3,000 a month in England.

31. CBS is proposing this service with the teletext system it is developing. Since several different teletext technologies are available; the industry in the United States is currently enmeshed in a debate over technical standards. The Electronic Industries Association (EIA) subcommittee on teletext, which is charged with selecting one of the three competing systems and submitting a set of standards to the FCC, has seen CBS drop out from the committee in order for the network to present its own standards petition to the FCC. A majority of the committee favors the British Ceefax system, CBS favors the French Antiope system. BROADCASTING, August 11, 1980, at 26.

32. This waiting requirement tends to limit the number of frames in a transmission cycle. In the case of Ceefax, 100 frames are transmitted in 25 seconds. A larger-frame memory capacity in the in-set decoder may mean that waiting time would not be a problem in the future. CBS is predicting an in-set memory capacity of up to 25 frames when its system starts up in 1985/86. BROADCASTING, August 6, 1980, 62, at 63.

33. The Prestel database is reported at 150,00 frames of information, with regional computer centers holding data in five cities. Prestel's service, which has been commercially available since 1978, now reaches half of Britain's telephone subscribers. Despite these seemingly impressive figures, there are only 2,400 Prestel equipped sets in use in Britain, and many of these are test or do-

nated sets. 8 INTERMEDIA, May, 1980, 13, at 4. Set supply has been a problem. See, for example, Rimmer, *supra* note 16, at 38, but there are also more fundamental marketing problems, such as whether a demand for the service actually exists. See Rimmer at 32, Tyler, *supra* note 24, at 276.

34. Fedida, *The Viewdata Computer: How Information Is Retrieved at the Command of the Subscriber*, 84 WIRELESS WORLD 44, April 1978.

35. Grundfest and Baer, "Regulatory Barriers to Home Information Services," (hereinafter Grundfest & Baer), at 325, in Dordick, *supra* note 24.

36. The Arbitron media survey service puts cable penetration in the United States at almost 19 percent of TV households: 14,261,200 cable households out of 75,793,500 TV households. Fourteen U.S. markets have penetration rates of 60 percent or more, with Palm Springs, California, highest at 99 percent penetration. BROADCASTING, February 18, 1980, at 98. An industry spokesman predicts a penetration rate for cable of 40 to 50 percent of TV households by the mid-1980s, "We're All in This Together Now," interview with Thomas E. Wheeler, BROADCASTING 32, March 3, 1980 (hereinafter Wheeler) at 32. Wheeler is president, National Cable Television Association (NCTA).

37. Grundfest and Baer, *supra* note 35, at 326.

38. Ashe, *supra* note 19.

39. Zerbinos, *supra* note 6.

40. 47 C.F.R. § § 76.205, 209 (1973). The FCC has applied the Fairness Doctrine and equal opportunities in political broadcasting to cable TV in the same manner that the doctrines are applied to broadcasting. For an extended discussion of the Fairness Doctrine and cable TV, see Barrow, *The Fairness Doctrine: A Double Standard for Electronic and Print Media*, 26 HASTINGS L.J. 659, 691 (1975). Watts, *supra* note 12. at 8, notes that the American Newspaper Publishers Association (ANPA) successfully petitioned for and obtained a decision from the FCC that its fairness rules would not extend to a videotex service delivered over a cable system. That is perhaps a more liberal interpretation of the holding in that petition than the FCC intended. *In re* the Matter of Amendment of Part 74, Subpart K, of the Commission's Rules and Regulations Relative to Community Antenna Television Systems; And Inquiry into the Development of Communications Technology and Services to Formulate Regulatory Policy and Rule-making and/or Legislative Proposals, Report and Order, 23 F.C.C. 2d 825, 829 (1970), when speaking of the Fairness Doctrine and equal opportunities rules as applied to cable, the FCC, at 820, said, "We did not intend to apply these requirements to the distribution of printed newspapers to their subscribers by way of cable . . . . We have no intention of regulating the print medium when it is distributed in facsimile by cable, but we do hold that the publication of a newspaper by a party does not put it in a different position from other persons when it sponsors or arranges for the presentation of a CATV origination which does not constitute the distribution of its newspaper." The idea of a facsimile newspaper involved the installation of a printer in the home which would output (usually at the publisher's control rather than the user's) an exact copy of that day's newspaper. This is a far cry from the electronic newspaper this study addresses, where the news product is

tailored to a user specified "profile" (*supra* note 4) of subject interest, which will be offered on a TV screen and which may or may not be offered as a printed copy, let alone as an exact facsimile of that newspaper. A tailored news profile would probably not "constitute the distribution of . . . newspaper," but rather would be a "CATV origination," at 830. For a discussion of early facsimile trials see Koehler, *Facsimile Newspapers: Foolishness or Foresight?* 46 JOURNALISM.

41. See, for example, Watts, *supra* note 12, at 6.

42. Scarcity and other rationales for broadcast regulation are considered at notes, and accompanying text *infra*.

43. Barnouw, THE GOLDEN WEB 18–22 (1968), notes the lack of interest in news and public affairs by the early broadcasters; Smith, Davis and Shelby, "Broadcast Executives' Attitudes Towards Fairness, Equal Time, Ascertainment, and Communications Act Revision" (paper presented to the Association for Education in Journalism Annual Convention, Boston 1980), at 15, found in a study of broadcast executives, in the context of a Communications Act rewrite, that the deregulation fought for by the industry on First Amendment grounds was not expected to result in increased news and public affairs programming. In fact, a majority of the respondents (66%) agreed that some stations would actually become less diverse if, for example, the fairness requirements were eliminated. Lucoff, *The Rise and Fall of the Third Rewrite*, 30 JOURNAL OF COMMUNICATION 47 (1980) (hereinafter Lucoff, Third Rewrite), notes at 52 that "pocketbook issues outweigh freedom of expression considerations."

44. An extreme example of a multimedia mix in one system is that of the security system offered by Warner Amex's Qube. The system is connected by cable to sensors in the home. If the cable or central computer is inoperative, a backup system cuts the alarm system onto the telephone system. *Warner Amex Plan Security System for Business, Home*, WALL STREET JOURNAL, March 12, 1980, at 10. (Less than a week later, Warner Amex announced that more than 700 families had ordered the security system connection. BROADCASTING, March 17, 1980 at 87). As an add-on to the telephone backup noted here, it is conceivable that an over-the-air "backup to the backup" could be offered. Although this example involves a security system and not an electronic publisher, it is also conceivable that an electronic publisher might involve more than one medium in its service with a cable down line delivering the information service and the telephone used to deliver the user's return requests.

45. In regard the matter of Amendment of Section 64.702 of the Commission's Rules and Regulations (Second Computer Inquiry), Report and Order, 77 F.C.C. 2d 384, 1980 (hereinafter Second Computer Inquiry).

46. FCC v. *Midwest Video Corporation*, 99 S.Ct. 1435 (1979) (Midwest Video II).

47. *National Association of Regulatory Utility Commissioners* v. *FCC*, 533 F. 2d 601 (D.C. Cir. 1976) (hereinafter NARUC II).

48. For example, H.R. 3333, 96th Cong., 1st sess. (1979), *Communications Act of 1979*, introduced March 29, 1979, lapsed July 13, 1979; H.R. 6121, 96th

Cong., 1st sess. (1979), *Telecommunications Act of 1979*, introduced December 13, 1979; S. 2827, 96th Cong., 1st sess. (1980), *Communications. Act Amendments of 1980*, introduced June 12, 1980.

49. Temple, *Technology Meets Bureacracy: The FCC's Policy for Two-Way Television*, 31 Fed. Com. L.J. 806 (1979), at 458 (hereinafter FCC's Two-Way Policy).

50. Telecommunication service is defined as "the offering for hire of a telecommunications capability for the transmission of information selected by the customer from one location to another by means of electromagnetic transmission with or without benefit of any physical transmission medium, including all instrumentalities, facilities, apparatus and services..." Principles of Agreement (between USITA and NCTA), June 19, 1980, at 1. I am indebted to Robert H. Glaser, assistant vice president, public affairs, Southwestern Bell for a copy of this agreement and for more general information about AT&T interest in electronic publishing. Telephone conversation September 18, 1980, and correspondence October 8, 1980 (hereinafter Glaser).

51. Wheeler, NCTA, *supra* note 36, at 34.

52. The Communications Act of 1979: Hearings on H.R. 3333 Before the Subcomm. on Communications of the Comm. on Interstate and Foreign Commerce, 96th Cong. 1st sess. 1270 (1979) (statement of Weldon W. Case, May 3, 1979). Case is the treasurer of U.S. Independent Telephone Association (USITA). USITA represents over 1500 (95%) of the non-Bell independent telephone companies in the United States and 97% of that industry in terms of gross revenues. Its largest member is GTE.

53. The Communications Act of 1979: Hearing on H.R. 3333 Before the Subcomm. on Communications of the Comm. on Interstate and Foreign Commerce, 96th Cong. 1st sess. 497 (1979) (statement of Charles A. Brown, April 26th, 1979). Brown is the chairman of AT&T. Since Brown's testimony, there has been no shift in AT&T position on this point of intrusion on traditional CATV activity. Glaser, *supra* note 50. See also *Promises, Promises*, Broadcasting, March 16, 1981, at 56.

54. *Supra* note 31. Broad, *Upstart Television: Postponing a Threat*, 210 Science 611, November 1980, at 614, suggests that CBS, in opting out of the EIA standards setting process for teletext, was actually working a delay tactic against the new medium to protect its broadcast advertising revenues.

55. *CBS Going Live with Teletext in LA Test*, Broadcasting, November 17, 1980, at 21.

56. Wollert, *Technology and Journalism: What Will the Future Hold?* 60 Forum 9, 10 (1980) (hereinafter Wollert, Technology and Journalism).

57. All the News That's Fit to Compute, *supra* note 17, at 18.

58. See, generally, Lucoff, Third Rewrite, *supra* note 43.

59. *Supra* notes 50 and 53 and accompanying text.

60. Sargent, *The First Amendment: Is It Being Threatened by Computer Information Systems?* ANPE Bulletin, November 1979 (hereinafter Sargent, First Amendment Threatened by Computer), at 3.

61. *Supra* note 17.

62. Ris, *Electronic Newspapers Could Alter Shape of the $4.6 Billion Classified Ad Market*, WALL STREET JOURNAL, August 11, 1980, at 13 (hereinafter $4.6 Billion Classified Ad). This article notes that AT&T offered a videotex service in Albany, New York, for six months early in 1980. Information made available consisted of white and yellow pages. minus display ads, from about 40 telephone directories in the Albany, Schenectady, Troy areas plus the directory for Manhattan. AT&T is proposing a test service in Austin, Texas, in mid-1981. See Hight, *Ma Bell Chooses Austin for Experiment*, AUSTIN AMERICAN-STATESMAN, September 11, 1980, at B3 (hereinafter, Ma Bell Chooses Austin). The extreme example of telephone directory automation comes from France where the state-owned telephone utility is proposing to supply all its customers with small black and white TV sets connected to computer-maintained telephone directories. The advantages of a constantly updated facility plus a cost saving in printing and distributing paperbased directories is considered sufficient to justify the scheme. See *The Curious Assemble for Viewdata-80*, BROADCASTING, April 7, 1980, at 34.

63. $4.6 Billion Classified Ad, *supra* note 62.

64. This service is available in the AT&T/Knight-Ridder Viewtron test in Coral Gables, Florida, *supra* note 17.

65. Viewtron Test Started by Knight-Ridder, *supra* note 17; Ma Bell Chooses Austin, *supra* note 62.

66. Viewtron Test Started by Knight-Ridder, *supra* note 17.

67. Ma Bell Chooses Austin, *supra* note 62. This Electronic Information Service, as it is called by the Bell spokesman, will involve 680 residences and 80 businesses and the trial will run for some 14 months.

68. See, for example, Sargent, First Amendment Threatened by Computer, *supra* note 60, at 4.

69. *Regulatory & Policy Problems Presented by the Interdependence of Computer & Communications Services & Facilities*, 28 F.C.C. 2d 291 (1970) (hereinafter Tentative Decision); 28 F.C.C. 2d 267 (1971) (hereinafter Final Decision).

70. 47 C.F.R. § 64.702 (1971)

71. The separate data processing entity was required to have separate books of accounts, separate officers and operating personnel, and separate equipment and facilities devoted exclusively to data processing services. The carrier was further prohibited from promoting the separate data processing service. Second Computer Inquiry, *supra* note 45, 391 n.2.

72. Second Computer Inquiry, *supra* note 45, paragraph 20 at 391; see also *re* the Matter of Amendment of Section 64.702 of the Commission's Rules and Regulations (Computer Inquiry), Supplemental Notice of Inquiry and Enlargement of Proposed Rulemaking, 64 F.C.C. 2d 771 (1977), paragraph 2.

73. For example, a three-class system of voice, basic nonvoice (BNV), and enhanced nonvoice (ENV) proposed in the Tentative Decision, *supra* note 69, paragraphs 8–58, was then discarded because of uncertainty about the nature of the service and whether the maximum subsidiary separation rule applies; *supra* note 71. The enhanced nonvoice category apparently could include both

regulated communication service and unregulated data processing. This required further development of the definitions of what each service was. But the gray line between the two would not go away, and the commission conceded that attempts to draw regulatory boundaries on the basis of precise technical distinctions would be rendered obsolete by technical development.

74. The basic transmission service is defined as "one that is limited to the common carrier offering of transmission capacity for the movement of information. In offering this capacity a communication path is provided for the analog or digital service depending on the following factors: (a) the bandwidth desired, (b) the analog and/or digital capabilities of the transmission medium, (c) the fidelity, distortion, or other conditioning parameters of the communication channel to achieve a specified transmission quality, and (d) the amount of transmission delay acceptable to the user." See Second Computer Inquiry, *supra* note 45, 419, paragraph 93. An enhanced service is "any offering over the telecommunications network which is more than a basic transmission service . . . In these services additional, different or restructured information may be provided the subscriber through various processing applications performed on the transmitted information, or other actions can be taken by either the vendor or the subscriber, based on the content of the information transmitted through editing, formatting, etc." Second Computer Inquiry, *supra* note 45, 421, paragraph 97.

75. *Id.* at 421, paragraph 97.

76. *Id.* at 433, n. 44.

77. *Id.* at 442, paragraph 149.

78. *Id.* 469, paragraph 222; "Inherent in the resale structure is the fact that the separate corporate entity may not construct, own, or operate its own transmission facilities." *Id.* 474, paragraph 229.

79. The Final Decision of the Second Computer Inquiry included GTE as a dominant carrier which should offer enhanced services through a separate subsidiary. In hearings to reconsider (adopted October. 28, 1980), the FCC excluded "GTE from the separate subsidiary requirement, leaving AT&T as the only dominant carrier subject to the subsidiary requirement." See Second Computer Inquiry, 48 RR 2d 1107 (1980), at 1122, para. 66 (hereinafter Reconsideration Order). The deadline for structural separation of enhanced services is March 1, 1982. *Re* Amendment of Section 64.702 of the Commission's Rules and Regulations (Second Computer Inquiry), Memorandum Opinion and Order, 79 F.C.C. 2d 953 (1980) 956 at para. 4. AT&T is reorganizing itself to set up this separated subsidiary. See BROADCASTING, August 25, (1980), at 11.

80. It would appear that AT&T will continue its plans for the proposed Austin trial, in spite of the rules ordered in the Second Computer Inquiry, for the following reasons: (a) There is a vacuum concerning the impact of the Inquiry pending resolution of challenges to the rules; 33 petitions to intervene have been filed at the U.S. Court of Appeals (D.C. Cir.), by parties claiming that the FCC does not have the authority to deregulate under the exisitng statute, that it misinterpreted the 1956 Consent Decree, and/or that it exceeded its authority by reserving the right to regulate data processing and data com-

munications, if necessary. REPORT together with DISSENTING VIEWS ON THE TELECOMMUNICATIONS ACT OF 1980 (H.R. 6121) BY THE COMMITTEE ON INTERSTATE AND FOREIGN COMMERCE, 96th Cong. 2d sess. August 25, 1980, at 46 (hereinafter Interstate and Foreign Commerce Committee Report). (b) The failure of the Communications Act rewrites in the Congress has set up another vacuum in authority. See, for example, note 83 *infra* and accompanying text. (c) There is an apparent loophole in the rules and legislation concerning the FCC's jurisdiction in intrastate v. interstate regulation (see *infra* notes 92–100 and accompanying text on NARUC II). A local service may not trigger the rules. (d) The Austin plan is a restricted trial, not a service offering. AT&T and Southwestern Bell's trial plans in Austin have been further complicated by suits to deny the telephone company the right to run the trial. The suits were initiated by the Texas Daily Newspaper Association (TDNA) and are being heard before the Public Utility Commission of Texas, *Re* Docket No. 3617 Complaint of Texas Daily Newspaper Association Concerning Expansion of Operating Authority of Southwestern Bell Telephone Company Beyond the Rendition of Telecommunication Service, (1981), with appeals in the District Court of Travis County, Texas 147 Judicial District, No. 318, 373 *Southwestern Bell Telephone Company* v. *Public Utility Commission of Texas* (1981); and in the Court of Civil Appeals, 3rd Supreme Judicial District, Austin, Texas, Case No. 13,471 *Southwestern Bell Telephone Company* v. *Public Utility Commission of Texas* (1981). (e) AT&T reorganized subsidiary (*supra* note 79) will not be ready to operate for some time, yet the company feels the need to begin to acquire experience with information services now. See Glaser, *supra* note 50. AT&T interprets the Reconsideration Order as allowing development of new enhanced services *prior* to the establishment of the separate subsidiary provided that the costs incurred are reported to the FCC. *In Re* PUC Docket No. 3617, Southwestern Bell's Reply to Exceptions of Texas Daily Newspaper Association, February 13, (1981), at p. 3.

81. *United States* v. *Western Electric Co., Inc. et al.*, (D.N.J. 1956) paragraph 63.246 (hereinafter 1956 Consent Decree), along with related FCC rules, bars AT&T from offering consumers new services that are commercially and technically incidental to communication, such as data processing and information services. All AT&T was permitted to do was provide communication services and facilities, the charges for which are subject to regulation.

82. Second Computer Inquiry, *supra* note 45, at 491, paragraph 272.

83. ADVERSE REPORT together with ADDITIONAL AND SUPPLEMENTAL VIEWS ON TELECOMMUNICATIONS ACT OF 1980 (H.R. 6121) BY THE COMMITTEE ON THE JUDICIARY, 96th Cong. 2d sess. October 8, 1980. The committee recommended that the bill not pass. After H.R. 6121 was reported favorably by committee on Interstate and Foreign Commerce on August 25, 1980, the Speaker ordered the bill referred to the Committee on the Judiciary to permit consideration of antitrust issues addressed by the bill. At 1.

84. H.R. 6121, 96th Cong., 2 sess., § 218. (a) (2) (A) (9i). "The term 'mass media' includes but is not limited to, television and radio broadcasting, pay television, and printed or electronic publications (including newspapers

periodicals, and any service or product like or similar to all or part of the function of newspaper or periodical or any portion of a newspaper or periodical). Such term does not include telephone number or address listings and directory assistance (limited to telephone number, address, and business category), weather or time information, or data retrieval services which do not provide any information which is like or similar to information provided by newspapers, periodicals, or television and radio broadcasting. The Commission shall have authority to determine in disputed cases whether any proposed service or product is a mass media service or product. *Id.* § 218. (e) (3).

85. *Key Players React*, TELEPHONY, April 21, 1980, at 19.

86. See, generally, FCC's Two-Way Policy, *supra* note 49.

87. *In re* Amendment of Part 74, Subpart K, of the Commission's Rules and Regulations Relative to Community Antenna Television Systems, Cable Television Report and Order, 36 F.C.C. 2d 143 (1972) (hereinafter 1972 Cable Report). The FCC did not require two-way services, only that the technical capacity to provide these services be allowed for when and if they became economically feasible. This technical/operational dichotomy has become an issue in franchise renewal proceedings in Austin, Texas, where the present franchise holder is offering "two-way capability" throughout a plant rebuild. An independent consultant has pointed out in his report that "capability" may not be the same thing as "operational or activated capacity." See REPORT ON THE EVALUATION OF CAPITAL CABLE COMPANY FOR THE CITY OF AUSTIN, TEXAS, August 20, 1980 (report prepared by Cable Television Information Center, Washington, D.C.) at VII-2 (hereinafter Capital Cable Report).

88. *Midwest Video II, supra* note 46, at 1441.

89. *Id.* at 1445.

90. *Id.* at 1445, n. 18.

91. *Cox Cable Gets Omaha Franchise; Two-Way Proposal Is Winning Factor*, BROADCASTING, August 25, 1980, at 110; *Warner Amex Lands Big One: Pittsburgh (with its Qube system)*, BROADCASTING, February 4, 1980, at 37; the NCTA is now surveying and publicizing the local cable programming activities of its members, *NCTA Report on Local Cable Programming*, BROADCASTING, September 1, 1980, at 41.

92. *Supra* note 47.

93. NARUC II, *supra* note 47, at 613.

94. *Id.* at 616.

95. One writer suggests that although the court in *NARUC II* explicitly discussed only jurisdiction over point-to-point transmissions, its conclusions in the case would probably extend to other more complex examples. *FCC Lacks Jurisdiction Over Two-Way Non-Video Intrastate Communications on Cable Television Leased Access Channels—National Association of Regulatory Utility Commissioners v. FCC (D.C. Cir. February 10, 1976) (No. 75–1975)*. Recent Cases, 80 HARV. L. REV. 1257, 1259 (1976)

96. *NARUC II, supra* note 47, at 609.

97. *Id.*

98. *Id.* at 608.

99. *Id.* at 610. At 608, n. 27, the Court noted that the FCC had, in another context, argued successfully that it is the character of the communication, rather than the character of the facilities, which determines the exclusion under 47 U.S.C. § 152(b), *General Telephone Co. v. FCC*, 134 U.S. App. D.C. 116 127–28 n. 19, 413 F.2d 390, 401–2 n. 19, *cert. denied.* 396 U.S. 888 (1969). The character of the communication is also the standard under which the Second Computer Inquiry defined the categories of basic and enhanced transmission services. This intermixing of agency ideas about the character of the communication, and legislative language concerning the facilities used, will probably continue to confuse the situation for some time, until either new legislation is passed or the Second Computer Inquiry clears its numerous challenges.

100. 47 U.S.C. § 152 (b), 2(b), states that "nothing in this chapter shall be construed to apply or to give the Commission jurisdiction with respect to (1) charges, classifications, practises, services, facilities, or regulations for or in connection with intrastate communication service by wire or radio or any carrier..." The FCC's reaction to *NARUC II* was that the point decided in the case was a narrow one which would not foreclose the Commission's authority to require that cable systems construct with the capacity to provide two-way services. The FCC suggested that some of the important services which two-way capacity makes possible, e.g., operational monitoring of the system's functioning, are related to the distribution of broadcast programming and thus meet the "reasonably ancillary standards" (*infra*, note 102) necessary for FCC jurisdiction. In re Amendment of Part 76 of the Commission's Rules and Regulations Concerning the Cable Television Channel Capacity and Access Channel Requirements of Section 76.251, Report and Order, 59 F.C.C. 2d 294 (1976) at 310.

101. Quoted in *Midwest Video Corp. v. FCC*, 571 F.2d 1025, 1053 (1978).

102. The "'reasonably ancillary' to the FCC's responsibility for broadcasting" standard, which was devised as a means to regulate cable television, received the Supreme Court's blessing in *U.S. v. Southwestern Cable*, 392 U.S. 157 (1968), and in *U.S. v. Midwest Video Corp.*, 406 U.S. 649 (1972) (hereinafter Midwest Video I).

103. See, generally, Emerson, THE SYSTEM OF FREEDOM OF EXPRESSION (1970).

104. Silber, BROADCAST REGULATION AND THE FIRST AMENDMENT (Journalism Monographs No. 70, November 1980).

105. *Columbia Broadcasting System, Inc. v. Democratic National Committee,*, 412 U.S. 94 (1973).

106. *FCC v. Pacifica Foundation*, 438 U.S. 726, 748 (1978).

107. *Id.* at 749.

108. See, generally, Powe, Background paper, The Edward R. Murrow Symposium on Press Responsibilities and Broadcast Freedoms, 17 (n.d., 1979?); Bazelon, *The First Amendment and "New Media"—New Directions in Regulating Telecommunications*, 31 FED. COMM. L.J. 201 (1979) (hereinafter Bazelon). Bazelon's thesis is that the new media have nullified the scarcity theory as a

basis for regulation, but that an "impact" (read, "power"?) theory has supplanted scarcity as a basis for regulation, at 207.

109. Bazelon *supra* note 108, at 209; See also Bazelon's dissent in *Brandywine-Main Line Radio, Inc.* v. *FCC*, 473 F.2d 16, 76 (D.C. Cir. 1972): Hagelin, *The First Amendment Stake in New Technology: The Broadcast-Cable Controversy*, 44 CIN. L. REV. 426 (1975). When the Supreme Court denied the FCC its content-oriented access regulations in *Midwest Video II*, it also denied one of Hagelin's "technological solutions," since the two-way requirements had been grouped with the access rules.

110. *NTIA Wants to Lift Fairness Obligations from Cable Systems with Access Channels*, BROADCASTING, May 26, 1980.

111. *Freedom of the Press*, BROADCASTING, May 19, 1980, at 79.

112. *Home Box Office, Inc* v. *FCC*, 567 F.2d 9, 46 (D.C. Cir. 1977), *cert. denied*, 434 U.S. 829 (1978).

113. See, for example, address by Charles Ferris, "Broadcasters and the First Amendment: No Stone Tablets," to the Federal Communications Bar Association, Washington, D.C. (November 30, 1979).

114. Busterna, *Ownership, CATV, and Expenditures for Local Television News*, 57 JOURNALISM QUARTERLY 287 (1980), at 290.

115. See, for example, *NTIA Comments and Recommendations on Communications Common Carrier Legislation*, Interstate and Foreign Commerce Committee Report, *supra* note 80, at 140.

116. Kaplan, "Trends in Persuasion on the Media: The Electronic Salesman and its Relationship to the Viewer" (paper presented at the Conference of the World Future Society, Toronto, 1980), at 10.

Chapter Nine

Technology Assessment for the Information Society

Jennifer Daryl Slack

Policy making for the information age is in turmoil.[1] Policy instruments constructed to serve political and economic exigencies of the past no longer seem adequate to support and serve the political and economic configurations taking shape in the information age. Clearly, new polices are called for, policies that take into consideration the uniqueness of the new technologies and the changing political and economic environment. In formulating and evaluating policy, however, we must remain sensitive to and critical of the political, economic, and ideological commitments embodied in the policy. Policy making is never an objective, value-free process. From the definition of the problem, through the characterization of the parameters of an acceptable solution, to the proposal for actual intervention in the political economic environment, policy making is a value-laden process.

Central to any formulation of evaluation of policy for the information age is a conception of the relationship between technologies and society. In a very real sense, what we "do" about technology depends on what we think it is, what we understand to be its role in society, and how we assess that role. In this chapter I examine a particular research tradition, technology assessment (TA), which has been developed specifically to aid the policy-making process. In particular I evaluate TA as a tool for assessing the relationship between technology and society and as a guide for generating meaningful public policy. In the process of this evaluation, the political economic and ideological commitments of TA are revealed. These commitments, I believe, render TA an inadequate tool in the policy-making process.

The rationale for choosing TA as the object of this analysis is threefold. First, TA is more or less the official governmental method of assessing the impact of technologies for the purpose of congressional policy making. Consequently, it demands evaluation as a potential tool for guiding policy making for the information age. However, TA has historically not had a recognizably significant impact in the congressional policy arena. TA, as a research methodology, as developed largely with governmental support, has had considerable impact in two additional areas, the import of which constitute further rationale for analyzing and critiquing TA. Second, TA as a research methodology, rather than simply a congressional capacity, has had a considerable impact on the way in which governmental agencies and private and public institutions envision and study the relationship between technologies and society and intervene in that relationship. Third, scholars are increasingly relying on TA as a model for understanding the relationship between communication technologies and society. Ithiel de Sola Pool, for example, in *Forecasting the Telephone*, professes to perform a "retrospective technology assessment" to examine the impacts of the telephone and to tune up the model of assessment for the purpose of assessing the impacts of new technologies.[2] For these reasons then, for the fact that TA officially embodies what has unofficially become a pervasive way of understanding the relationship between technology and society and for providing a basis for intervention in that relationship, this analysis is intructive beyond the narrow conception of TA as a strictly congressional capacity.

The History of Technology Assessment

To assess technology, "to analyze critically and judge definitively the nature, significance, status, or merit of"[3] technology, is not really to do anything particularly new. Assessment of technologies has a long history, although such assessment was not, and often still is not, carried on under the official aegis of TA. Individuals or institutions often assess the merits of exploiting or not exploiting, developing or not developing, technologies in their control. Consumers assess a technology when, in the marketplace, they choose whether to buy. Conscious and deliberate assessment of technology has been conducted by government for some time, but historically this has been executed primarily after the fact, that is, after serious damage has occurred or after there has been public outcry. For example, it was not until there was a tremendous outcry by concerned parents, teachers, and social critics about the deleterious effects of television on children's eating, reading, and exercising habits that the government became involved in assessing the effects of television on children.

While the process of assessment—be it conscious and deliberate or merely subconscious—includes the measurement of merit, significance, or status, the criteria for measuring these attributes can, and do, vary widely, depending on the assessor, the purpose for which the technology is being assessed, and the socioeconomic-political situation in which the assessment is taking place. The realization of short- or long-term profit; the realization of use-value; the extension, expansion, or solidification of market control; the attainment of status; and service to another's interest (such as society's or another individual or institution's) are all examples of various measures by which merit or significance might be gauged. But in order to portray the nature of the technology, both the positive and negative aspects of technology must be measured. Therefore, demerits of the technology must be assessed as well.

Both the measurement of merit and demerit can involve the assessment of a past or current situation as well as the projection of a future situation. The latter gains prominence in the practice of assessing for the purpose of deciding to exploit or develop a technology. In this situation, after the potential merits and demerits, gains and losses, are projected, as well as, perhaps, the relative merits of various alternatives, judgment is then passed and a decision made that is expected to produce the greatest benefit, in accordance with whatever the particular expectations are in a particular case.

However, decisions in real life are not always as rational as the process outlined above might suggest. Often decisions are made that are convenient, ignorant, spiteful, or even coerced. Besides, a decision that may seem rational to one observer may seem irrational to another observer. As a result, therefore, of irrational decision making on the part of the decision maker and the differences in perceptions regarding the nature of rationality, decisions can appear to be, and may in fact be, haphazard, devoid of rationale, and even in opposition to the best interests of both the decision makers and those affected by the decisions.

As the concern over possible deleterious and irreversible effects of technology intensified in the late 1960s, the "mood" of many in government in Washington turned progressively toward a cautious skepticism of unrestrained technological development. It is in this mood of cautious skepticism, and in response to the knowledge that all too often technological decisions were based purely on self-interest or on the less "rational" criteria, decisions which may in fact have considerable impact on the survival of our nation (not to mention our world), that a movement began that was interested in developing a rigorous, institutionalized, rational congressional TA capability. The motivations of the people involved in this movement were varied.[3] Apparently, some participants were really anti-technologists desiring a legitimate way to suppress technologies. This group, however, does not seem to have played a very significant role in the definition of TA. The group that did play the most significant role in shaping TA was made up of "good government reformers," or "rational policy advocates." This group was decidely pro-technology, concerned with enhancing technological development while at the same time minimizing its undesirable effects.

The credit for giving the imprimatur to both the terminology and the current meaningfulness of TA is probably best ascribed to the efforts of Emilio Q. Daddario.[4] In 1967, while chairman of the Subcommittee on Science, Research, and Development of the House Committee on Science and Astronautics, Connecticut congressman Daddario introduced a bill calling on the federal government to establish a TA Board "to provide a method for identifying, assessing, publicizing, and dealing with the implications and effects of applied research and technology."[5] Daddario believed quite strongly that "technological changes have become so extreme and occur so rapidly that it is incumbent upon us to reverse the process."[6] In addition to the hope that a TA Board might be able to "anticipate and minimize the unwanted side effects which so often accompany innovation," Daddario was hopeful that a Board might be able to streamline or rationalize technical development, particularly given the high cost of modern research and development. "A Technology Assessment capability for Congress," Daddario asserted, "will enable us to deploy the finite scientific and engineering resources of money, facilities and skilled manpower to take fullest advantage of the gains offered to society."[7]

Daddario's bill was not passed in the late 1960s. By the early '70s however, there was a flurry of TA activities conducted within governmental agencies or with government support. In particular, the

National Science Foundation began funding TA projects, and a number of federal agencies claimed to be involved in conducting TA-type studies.[8] Finally, the Office of Technology Assessment (OTA) was created by Congress in 1972, pursuant to the Technology Assessment Act of 1972. The OTA received funding in November 1973 and began operating in January 1974. Daddario became its first director and remained in that position until May 18, 1977. Its basic mandate was to "provide Congressional committees with assessments or studies which identify the range of probable consequences, social as well as physical, of policy alternatives affecting the uses of technology."[9] Six priority areas, designated by the OTA's Congressional Board, were adopted by the OTA: oceans, transportation, energy, materials, food, and health.[10] These priority areas were later structured into three major divisions: energy, materials, and global security; health and life sciences; and science, information and transportation.[11]

Until 1978 most assessments were conducted in response to specific requests from congressional committee chairpersons or ranking minority members.[12] An effort was made in 1978 to begin to assess technologies based on an internally drawn up "priority list of issues of critical concern to the United States and the world."[13] The final decision to assess a technology would depend on the possibility that it might have significant impact, that the assessment might provide foresight, that there was congressional interest in the issue, and that the OTA was capable of performing the task.[14]

The approaching information society was not initially a priority area of interest to the OTA. A few studies of telecommunications were conducted before 1978, the first being an assessment of the value of broadband communication systems in rural areas.[15] In 1978, however, the OTA began addressing the problems of telecommunication systems in a more rigorous way, by establishing the Telecommunications and Information Systems Group. The reason that the OTA offered for this new concern was that due to the rapid advance of new technologies in communication," new legislation is being proposed and adopted, and relevant international norms are being formulated."[16] Furthermore, the OTA had been asked by several congressional committees to assess the situation. Two assessment projects specifically concerned with telecommunication were begun in 1978, the first assessing the societal impacts of national information systems and the second the impact of new telecommunication technologies such as satellites and fiber optics.[17] The First of these, *Computer-Based National Information Systems*, was completed in late 1981.[18]

The Theory and Practice of Technology Assessment

It is tempting to try to separate the theory of TA from the practice of TA, for there is considerable debate, discussion, and argument about what TA ought to do or could do—and then there is what TA does do.[19] But despite some differences between what TA could or should do and what it does, both the theory and practice of TA seem to be grounded in some crucial assumptions, assumptions of considerable consequence for TA's ability to interpret the relationship between technology and society and to provide a basis for meaningful public policy.

I think it is possible to summarize my concerns about TA by asserting that TA is not really about assessing technologies, that is, not in the original sense of "to assess." Recall the dictionary definition of assess quoted earlier (a definition Daddario used frequently): "to analyze critically and judge definitively the nature, significance, status, and merit of." Now compare this definition of assess with a definition of TA that has been very widely accepted as an acceptable definition of institutionalized assessment: "the systematic study of the effects on society that may occur when a technology is introduced, extended, or modified, with special emphasis on the impacts that are unintended, indirect, and delayed."[20] There are quite a few differences between these two definitions, but two are particularly significant vis-á-vis the theory and practice of TA. First there is a shift from assessing a whole technology to the study of the effects of a technology. Second, the study of effects is still further limited to the emphasis on unintended, indirect, and delayed impacts. What of the intended, direct, and immediate impacts? An examination of these two emphases in TA reveals that TA is an inadequate model for the analysis of the relationship between telecommunications and the information society and for intervention in that relationship.

TA focuses on the impact or effects of technologies. Analysis usually begins with the identification of impact areas, first macro areas (such as the economy) and then micro areas (such as the GNP or employment). Researchers then identify the specific impacts in these areas that characterize the particular social situation within which the technology has effects. So, for example, a causal chain might be constructed such that the computer causes less jobs for clerks, causing unemployment, causing retraining, causing employment. Different studies will bound (limit the length of) these causal chains depending on a number of factors, such as the purpose of the study, the methodology employed, or the concern for limiting the extension of the chain to the point where the researchers feel certain that they are still talking about the effects of the technology.

This characterization of the technology as having impacts on society is based on an inadequate conception of the causal relationship between technology and society. The conception is a mechanistic one, that is, one that conceives of causes and effects as discrete and isolated objects, events, or conditions that exercise effectivity externally. Both the cause and the effect are self-contained and distinct from their environments, as is the cause-effect event. The technology, as the cause, is an autonomous phenomenon that exercises effectivity on another phenomenon that is otherwise also autonomous. The world is unrealistically divided into simple, discrete parts, essentially unrelated, except perhaps for the moment when a cause produces an effect, after which the parts are again simple, discrete, and unrelated.

In a mechanistic position that can be termed "simple," neither the technology nor its impacts are inherently linked to the environment within which the technology has arisen and within which it has effects. The technology is isolatable; it causes the effects all by itself, and the effects are inevitable. So, for example, in an information society the computer might be seen as necessarily causing less jobs for clerks, which necessarily causes unemployment, and so on.

In a mechanistic position that can be termed "symptomatic," the technology is likewise a discrete phenomenon with no essential connectedness to the environment within which it arises, but it is envisioned as entering into a system where institutional, social, and cultural forces interact with it to shape its effects. So, for example, the computer will enter into a certain social structure where it will be put to use in certain ways; the effects of the computer will be dependent on the particular uses to which it is put. From this perspective, the computer might cause unemployment if adequate retraining programs are not also instituted. The computer is still envisioned as having effects; it is just that those effects can be mitigated or shaped by other social forces. TA tends to conform to a symptomatic model of causality, for TA is predicated on the belief that it can generate policy options that can both mitigate undesirable effects and enhance desirable ones.

The utilization of a symptomatic model of causality has serious implications for the ability of TA to characterize adequately the relationship between telecommunications and the information society and to suggest effective intervention. TA has not been able to take into account the context within which telecommunication appeared in the first place. It has not been able to look at telecommunications as both cause and effect, as part of a social configuration in which effectivity is exercised as a relationship between the parts of a social structure. So, for example, the cause of unemployment can be traced only to the computer. It cannot be traced to the particular political, economic, and ideological configuration within which computers arise—already value laden—and within which they exercise effectivity as merely a part of the configuration. If the computer was searched for, invented, and innovated primarily as an efficient method of decreasing labor costs and enhancing efficiency, the "cause" of unemployment is every bit as much (if not more) attributable to the ideology of efficiency, economic strategies, contradictions between labor and capital, and the conjunctures between them. Appropriate intervention would have to be based on an assessment of the totality of those relationships within which the technology was a part. Only then would assessment and intervention really be based on the nature, significance, status and merit of the technology.

Confounding the problems of conceiving of the relationship between technologies and society as a mechanistic one in which technologies have effects, TA tends to limit the consideration of effects to those that are unintended, indirect, and delayed. As a consequence, it is unlikely that TA would ever acknowledge the primary motivations for which the technology in question might have been invented and innovated, let alone assess that motivation. TA does tend to affirm—if only tacitly— the primary functions of technologies as well as the intended motivations for implementation. TA thus tends to be not only a highly conservative activity but one that is aggressively so. TA professes to "assess" when what it really does is offer legitimation. To understand how this operates, it is important to consider the way in which TA equates technological development with progress.

The Politics of Technology Assessment: Technological Growth and Progress

In September 1967 the Subcommittee on Science, Research, and Development, under Daddario's chairmanship, sponsored a Seminar on Technology Assessment with the intent of beginning an exploration of the issues involved in TA as embraced by the bill submitted earlier. At the request of the House of Representatives, both the National Academy of Sciences and the National Academy of Engineering undertook to study TA and to submit reports of their findings. In particular, the two august academies were to explore the following aspects of TA: "what it means to various groups, how it occurs today, how it is related to the behavior of individuals and organizations, how its quality might be improved and its influence enhanced."[21] The reports submitted by the National Academy of Sciences (NAS)[22] and the National Academy of Engineering (NAE),[23] in addition to an article by Daddario published in the *George Washington Law Review* in 1968,[24] best characterized the intent and spirit of TA. Despite slight differences between them, all fundamentally ascribe to the same brand of faith in technological progress. Utilizing these three documents, I will elucidate that faith.

The impetus for the genesis of TA emerges as twofold. TA is at once a response to the negative consequences of technological development due to unplanned, anarchic freedom, as well as a response to the desire to fully exploit technologies—in the guise of progress. The latter requires the minimization of any public dissent and/or reluctance that might inhibit potential development. This twofold impulse is expressed in Daddario's desire for a TA board to streamline technical development and is similarly expressed in the "Summary of Findings" of the NAE report: "Unless dependable means are developed to identify, study and forecast the varying impacts that these technological developments might have on sectors of our society, the nation will be subjected to increasing stress in time of social turbulence and will not benefit fully from technological opportunities."[25]

In order to monitor undesirable social effects while fully exploiting potentials, the NAS argued that a rigorous standard for TA must be designed. The report recognized that assessment of technological prospects based primarily on self-interest has become widespread in both private and public institutions. One problem with such self-interested analysis, the NAS maintained, was that it "may ignore important implications of particular choices for sectors of society other than those represented in the initial decisions. In their pursuit of benefits for themselves or for the particular public they serve, those who make the relevant decisions may fail to exploit technological opportunities that, from a broader perspective, might clearly deserve exploitation. Likewise, as they seek to minimize costs to themselves, the same decision-makers may pursue technological paths that, again from a broader perspective, ought to be redirected so as to reduce undesirable consequences for others. A wide variety of what economists call external costs and benefits thus falls 'between the stools of innumerable individual decisions to develop individual technologies for individual purposes without explicit attention to what all these decisions add up to for society as a whole and for people as human beings."[26]

The goal of TA is to provide information that will allow policy-makers to correct deficiencies in the uses of technologies. Indeed, the NAS claimed that the object of their study was not technology per se, but human behavior and institutions. Their goal was "not to conceive ways to curb or restrain or otherwise 'fix' technology but rather to conceive ways to discover and repair the deficiencies in the process and institutions by which society puts the tools of science and technology to work."[27]

While TA enables one to generate limited criticism and analysis of the uses of particular technologies, the approach is still firmly committed to the credo that technology is essentially synonomous with progress. Nowhere is this more explicit than in the NAS report, where man is depicted as "committed to a highly technological culture." Yes, there are "technologies whose effects we sometimes deplore," but these very same technologies are responsible for liberating us: these are the very same technologies which "are themselves largely responsible for the fact that we both can and do consider the effects of decisions and policies on a much larger part of the human population than ever before." It is the very advances in technological development that have enabled us to "anticipate the secondary and tertiary consequences of contemplated technological developments and to select those technological paths best suited to the achievement of broad combinations of objectives." Technological development is credited with giving us more options from which to choose and, as a consequence, a superior ability to choose. As a result of this confidence, and in spite of the "long history of inattention to the wider consequences of technological change," the NAS report begins with, and is guided by, "the conviction that the advances of technology have yielded and still yield benefits that, on the whole, vastly outweigh all the injuries they have caused and continue to cause."[28]

Whatever TA does, it was felt that it must not be allowed to become a threat to technological progress. It must not, therefore, emphasize the dangers of proposed changes but their potential for good.[29] The NAS even suggested that the "burden of uncertainty" due to imperfect knowledge must not continue to be placed on those who develop new technologies and new uses for technologies.[30] We must still be willing to take chances, to experiment, as we always have been, for as Daddario stated, it is "the boldness to try something different...[that]... has been responsible for a great portion of our material welfare and strength among nations; it has set our country somewhat apart from other countries of similar culture."[31] The faith in technology and its ability to ensure our progress as a nation is still strong, albeit the notion of progress is a bit moderated. The challenge of technology assessment, according to the NAS, is "to discipline technological progress in order to make the most of this vast new opportunity."[32]

Technology Assessment as a Basis for Policy Making for the Information Age

As this analysis indicates, TA's conceptualization of the relationship between technologies and society is severely limited. It is, of course, necessary for any researcher to limit the scope of research, as it is impossible to answer all possible questions about any technology in one study or even in a myriad of studies. But it is also absolutely essential to understand and evaluate the ways in which any study—or in this case, a research tradition—limits the definition of the problem to be studied such that acceptable parameters for generating acceptable policy are also limited. TA so limits the scope of study as to promulgate an unquestionably uncritical and untenable bias toward technological growth and development and an equally uncritical and untenable equation of technological growth with social progress. Built into both the theory and practice of TA is the bias toward asking policy-oriented questions in a way that reaffirms the faith in technology as a tool to solve any and all social problems. This faith, firmly rooted in a symptomatic view of the relationship between technology and society, virtually assures that TA can never adequately assess the relationship between technology and society. TAs of telecommunications conducted by the OTA, as well as by public and private institutions and by individual scholars, have all been severely flawed due to the above commitments.[33]

In developing policy for the information age, we clearly do need to be able to assess telecommunications. But in order to maximize the likelihood that policy will be meaningful, assessment must be based on an adequate understanding of the relationship between telecommunications and society. Similarly, assessment must not be uncritically committed to the equation of technological growth with progress. Consequently, TA is an inappropriate model on which to base either an analysis of the technologies of the information society or the development of policies for the information society.

Notes

1. I thank Kenan Jarboe for his comments on an earlier version of this paper.

2. Ithiel de Sola Pool, *Forecasting the Telephone: A Retrospective Technology Assessment* (Norwood, N.J.: Ablex, forthcoming).

3. *Webster's Third New International Dictionary*, s.v. "assess."

3. Mark R. Berg, "The Politics of Technology Assessment," *Journal of the International Society for Technology Assessment*, 1 (December 1975), reprinted in *Science, Technology, and National Policy*, edited by Thomas J. Kuehn and Alan L. Porter (Ithaca: Cornell University Press), pp. 477–478.

4. Albert H. Teich, ed., *Technology And Man's Future*, 2d ed. (New York: St. Martin's Press, 1977), p. 223, also credits Daddario in this way.

5. U.S. Congress, House, H.R. 6698, 90th Cong., 1st Sess, 1967.

6. Emilio Q. Daddario, "Technology Assessment—A Legislative View," *George Washington Law Review*, 36 (1968), p. 1044.

7. Ibid., p. 1046.

8. Alan L. Porter and others, *A Guidebook for Technology Assessment Analysis* (New York: North Holland), pp. 34–37.

9. U.S., Congress, Office of Technology Assessment, *Annual Report to the Congress, 1974/1975* (Washington D.C.: Government Printing Office, 1975), p. 4.

10. Ibid., p. 15.

11. U.S., Congress, Office of Technology Assessment, *Annual Report to the Congress for 1978* (Washington D.C.: Government Printing Office, 1979).

12. Ibid., p. 4.

13. Ibid.

14. Ibid.

15. U.S., Congress, Office of Technology Assessment, *The Feasibility and Value of Broadband Communications in Rural Areas: A Preliminary Evaluation*, Staff Report, April 1976 (Washington D.C.: Government Printing Office, 1976).

16. U.S., Congress, Office of Technology Assessment, *Annual Report to the Congress for 1978*, pp. 55–56.

17. Ibid., pp. 56–57.

18. U.S., Congress, Office of Technology Assessment, *Computer-Based National Information Systems: Technology and Public Policy Issues*, Staff Report, September 1981 (Washington D.C.: Government Printing Office, 1981).

19. For a discussion of the range of debate within TA, see Mark Boroush, Kan Chen, and Alexander Christakis, *Technology Assessment: Creative Futures* (New York: North Holland, 1980), pp. 350–93.

20. Joseph Coates, "The Identification and Selection of Candidates and Priorities for Technology Assessment," *Technology Assessment* 2, no. 2 (February 1974): 77.

21. U.S., Congress, House, Committee on Science and Astronautics, *Technology: Processes of Assessment and Choice*, Report of the National Academy of Sciences, 91st Cong., 1st sess., July 1969 (Washington D.C.: Government Printing Office, 1969), p. 7.

22. Ibid.

23. U.S., Congress, House, Committee on Science and Astronautics, *A Study of Technology Assessment*, Report of the Committee on Public Engineering Policy of the National Academy of Engineering, 91st Cong., 1st sess., July 1969 (Washington D.C.: Government Printing Office, 1969).

24. See note 5.

25. *Technology: Processes of Assessment*, p. 3.

26. Ibid., pp. 9–10.

27. Ibid., p. 15.

28. Ibid., pp. 10–11.

29. Daddario, p. 1047.

30. *Technology: Processes of Assessment*, pp. 33–39.

31. Daddario, p. 1047.

32. *Technology: Processes of Assessment*, p. 12.

33. See, for example, the OTA's Broadband study (above); the OTA's National Information Study (above); Cary Hershey and Elizabeth Sachter, "Acquiring Baseline Data on Potential Uses of New Communication Technologies," *Journal of the International Society for Technology Assessment 2*, no. 1 (1976): 51–61; Edward M. Dickson and Raymond Bowers, *The Video Telephone* (NewYork: Praeger, 1974); Raymond Bowers, Alfred M. Lee, and Cary Hershey (eds)., *Communications for a Mobile Society: An Assessment of New Technology* (Beverly Hills: Sage, 1978); and de Sola Pool's retrospective assessment of the telephone (above).

Chapter Ten

The Telecommunications Revolution: Are We Up to the Challenge?

Jerry L. Salvaggio

The rapid emergence of telecommunication systems and their precipitous intrusion into all sectors of life has been greeted in this book and elsewhere with grave apprehension. The authors seem to agree that our cultural, social, and political patterns are destined to enter a period as different from our own as the industrial period was from the agricultural. To many of the contributors the information society is a telecommunications nightmare. Perhaps such pessimistic forecasts represent the intellectual and emotional remnants of industrial man and his wish for less radical change. Yet the chance that these prognostic warnings presage problems in the information society demands our attention.

What conclusions are to be drawn from these essays? Is there a singular concern that can be synthesized from an amalgamation of the views presented? If a singular apprehension has been expressed by those who have given much thought to the information society, it is the fear that telecommunications will result in information being misused and centrally controlled. Apprehension concerning the misuse and control of information by government, agencies, public utilities, and the information industry form a recurring motif in the literature on information societies with some believing these two activities to be indigenous to information societies.

The contributors to this book foresee a distinct threat to the individual in a society where telecommunications may be controlled by a few and all information may be centralized. Whom do the authors foresee as the major threat to individual liberty? Some would have technocracy as the tool of big business. Ernest Rose, for example, warns us that the information society may sustain the advantages of the corporate world while reducing protection of the individual. He notes that the rush to "decontrol" the telecommunications industry favors corporate property rights while ignoring individual rights. In an uncontrolled economic environment there is greater potential for lupine media giants to gain increased control of information. As Rose cautions: "It is disquieting to observe that neither government nor business have much of a track record in ethical behavior. Unfortunately, both will have great influence on information technology."

Herbert Schiller's essay echoes the sentiments of Rose, though he places greater emphasis on the fact that the corporate world is allied with government, which, Schiller notes, already has control over the technology. "In sum," he writes, "a great amount of the activity, a good share of the content, and the general thrust of what is now defined as the information age represent military and intelligence connections." Schiller contends, with Rose, that new telecommunication systems support the business component—more specifically, the transnational corporate community. Schiller, however, goes further than Rose by adding that information technology was designed by the military in order to eventually be the "ultimate enforcer."

What are the consequences if Rose and Schiller are correct in their contention that the corporate world and the government will take charge of a centralized telecommunications system? Almost all the prognosticators in this book touch upon the problem of invasion of privacy. Joseph Pelton conducted a Delphic survey that indicated most communication experts believe that by the year 2000, databases will be compiled on almost all people in information societies. Pelton reminds us that attempts to curtail invasion of privacy have not been very successful. Britain's Younger Commission and the National Commission for the Review of Federal Laws Relating to Wiretapping and Electronic Surveillance of the United States have failed to stymie the trend toward more invasion of privacy.

Collecting databases on individuals will certainly be a simple matter if the family is using an interactive information system such as inter-

active cable, a videotex system, or a personal computer tied into a
database. Interactive cable systems automatically scan each home subscriber every few seconds, tabulating data and monitoring the security and
fire alarms of each home. The intent is innocuous enough. The cable
company needs the information for billing and for monitoring security
systems. Where data on individuals are collected and stored, however,
there is the potential for misuse. The use of videotex systems for electronic mail delivery, teleshopping, and tabulating one's financial records will provide insurance companies, credit firms, and government
agencies with a tempting data base.

Invasion of privacy of a different sort will be facilitated in the information society through telemetering. One computer expert conjec-
tures that tiny sensors and transmitters similar to those being used to
keep track of heartbeats may be embedded in the human body. Initially, this may be done only with parolees so that their whereabouts can
be monitored. Later, whole classes of people representing a perceived
potential threat to society could be monitored.

Government and industry are not the only ones the contributions are
afraid of. Pelton explains that highly centralized computer systems
make us likely prey for technologically astute criminals, spies, and terrorists. If Pelton is correct, then the potential will exist for urban guer-
rillas with a different ideology from our own to inflict a catastrophe on
any information society. Embezzlement, credit card and magnetic card
forgeries, destruction of vital records, covert acquisition of secrets, and
falsification of data to incriminate or exonerate key officials, Pelton
notes, are conceivable in an information society.

If misuse of information and control of telecommunications are the
two most serious problems associated with the information society, the
major deterrent to our meeting the challenge, according to the authors
in this volume, is bureaucratic inadequacy. Thus the major concern of
those authors writing on public policy for the information era is the
lack of a U.S. public policy and the inadequacy of existing government
regulation of the information industry.

If anyone seriously believes that the United States has a long-range
public policy plan for using telecommunication, they need only
examine the situation in Japan to see that the United States lags at least
a decade behind in preparing for the information society. From the
account of Japan's plans for the twenty-first century provided by Ogasawara and Salvaggio, it is clear that a comprehensive public policy
must be based on a coordinated effort of the government, public utilities, and the information industry to meet a predetermined long-range
goal. Japan has not only managed to get industry and government

working together; its long-range goal is being tested in the form of multimillion-dollar experimental projects that re-create a miniature information society. In the United States the private sector has carried out experiments with videotex; and the Department of Commerce, in conjunction with the Department of Agriculture, has carried out experiments such as the Green Thumb project. But there is no coordinated effort to design national information systems other than those of the military. The United States has been slow to comprehend what the Japanese seem to have grasped: The economic and social bases of future societies will depend on efficient national information systems. There is little doubt, based on the information provided by Ogasawara and Salvaggio, that Japan will be in the forefront of information societies. Videotex, teletext, over-the-air universities, and electronic newspapers via direct broadcast satellite will be in operation on a national basis in Japan when the United States is still deciding on technical standards for videotex. Ogasawara and Salvaggio conclude that the MPT's organized and concerted effort, with the information society as a goal, might be an alternative model that America would do well to study.

If public policy and planning for the information era are lamentably absent, the situation in regard to regulating telecommunications is equally chaotic. William Read and Tony Rimmer point to the obsolescence of existing regulations in the area of journalism. Read contends that the information society will put some settled questions at issue once again, particularly "the judicial notion that a greater scope of government regulation is permissible as to broadcasting than as to the press." As Read and Rimmer point out, the notion is based on a two-track legal approach. Thus, publishers are protected by the First Amendment against government interference, while federal regulators continue to determine, for the broadcasters, how the public's best interest is served. Yet, as Rimmer observes, "the gathering and processing of information and getting it to a certain point down the channel of distribution will increasingly be the same. . . ." Both authors agree that this dualistic policy—one for print media, the other for broadcast media—has been technologically undermined and that if it is not changed, the major forms of news transmission (cable, videotex, teletex, etc.) in the information society will be legally controlled from the beginning. Read and Rimmer also agree that, if a choice must be made, it is the print standard that should prevail.

Another aspect of the regulatory problem relates to the increasing tendency to deregulate telecommunications. If this trend continues, the information society may be without a safeguard against those who gain control over the technology. Should they wish to misuse the information they have gathered, there would be little one could do. The problem is well put by Roland Homet: "As regulatory intervention fades, who or what will preside over the outcome? Will the struggle be left simply to survival of the fittest, or as Tennyson put it, to 'nature red in tooth and claw?' And where will be our social values then?" Homet contends that in an era of deregulation, private bargaining with the government will come to replace public decision-making. When that happens, Homet warns, "we are going to have to think carefully about the ramifications for democratic participation and control in our communication structure."

Ironically enough, it may not be the public that calls for increased regulation. Homet predicts that there will be pressure from certain aspects of the information industry itself for protective regulation to limit or prevent the convergence of communication services. The temptation to respond with more regulations, Homet agrues, is not the answer. Regulatory agencies too often suffer from human error and the facts available to them are all too often inadequate or out of date. The answer, for Homet, is market reputation, consumer resistance, and entrepreneurial inventiveness rather than government intervention.

The final problem the United States faces as it turns to meet the challenge of the telecommunication revolution is its inability to properly assess technology and its effects. Without an accurate assessment of the technology telecommunications, policy making is impossible. In 1972 Congress created the Office of Technology Assessment (OTA) for the purpose of identifying the range of probable consequences of policy alternatives affecting the uses of technology. As Jennifer Slack notes in her chapter, the approaching information society was not initially a priority area of the OTA. Since then, the OTA has begun a special focus on the unintended effects of communication technology. Slack, however, makes a strong case against their methodology and their initial assumptions, which, she contends, may prevent them from being successful. The OTA's conception of the causal relationship between technology and society is limited, according to Slack, in that the OTA takes a mechanistic position that unrealistically divides society into simple, discrete parts unrelated to the environment within which the technology has arisen. Slack further observes that technology assessment has a bias toward "asking policy-oriented questions in a way that reaffirms the faith in technology as a tool to solve any and all social problems." Slack thus concludes that technology assessment, as it is being conducted today, is an inappropriate model on which to base an analysis of telecommunications and its effects on society.

I have purposely saved Daniel Bell's chapter to discuss last, for it covers the question I consider to be the most important. If the tele-communication revolution can be expected to have adverse social effects, as the authors here have argued, what can be done? Bell's essay makes it clear that society can determine only whether the in-novations wrought by telecommunications will create a nightmare or a utopia. An elaborate telecommunications system, Bell maintains, allows for intensification of what in military parlance is called "com-mand and control systems." Yet Bell remarks further on, "by the very same technology, one can go in wholly different directions." "The rev-olution in telecommunications," Bell writes, "makes possible both an intense degree of centralization of power, if the society decides to use it in that way, and large decentralization because of the multiplicity, diversity, and cheapness of the modes of communication." In the poli-tical realm the technology makes a complete "plebiscitarian" system possible, allowing for each household to vote on a variety of issues. If conditions remain free, Bell adds, individuals could create their own modes of communication and their own new communities.

In summary, the revolution in telecommunications is a two-edged sword. Telecommunications could mean information societies with greater individual freedom, if society demands such freedom. A careful reading of these essays, however, indicates that the majority of the authors represented here are not optimistic.

Selected Bibliography

Abels, Sonia Leib; Paul Abels; and Samuel A. Richmond. "Ethics Shock: Technology, Life Styles and Future Practice." *Journal of Sociology and Social Welfare* 2, no. 2 (1974): 140–54.

Abrahamson, D. S., and R. P. Murray. Human Factors Study, Experiments 1 and 2: Keyboard Type & System Response Delay. Report on "Videotext: Human Factors Studies" submitted to Manitoba Telephone Service. August 1980.

Agostino, Donald E.; Herbert A. Jerry; and Rolland C. Johnson. "Home Video Recorders: Rights and Ratings." *Journal of Communication* 30, no. 4 (Autumn 1980): 28–35.

Alper, W. S., and T. R. Leidy. "The Impact of Information Transmission through Television." *Public Opinion Quarterly* 33 (1970): 556–62.

Alvarez, J. A.; W. W. Grote; and C. E. Nahabedian. "Conferences and Classes Via PCT: If You Can't Come, Call." *Bell Laboratories Record*, April 1973, pp. 99–103.

American Satellite Corporation. "American Satellite to Transmit Midwest Edition of the New York Times." *ASC Newsrelease*, April 23, 1980.

Androunas, Elena, and Yassen Zassoursky. "UNESCO's Mass Media Declaration: A Forum of Three Worlds." *Journal of Communication* 29, no. 2 (Spring 1979): 186–98.

Artandi, Susan. "Man, Information, and Society: New Patterns of Interaction." *Journal of the American Society for Information Science* 30, no. 1 (January 1979): 15–18.

Austing, Richard H. "A Study of Computer Impact on Society and Computer Literacy Courses and Materials." *Journal of Educational Technology Systems* 7, no. 3 (1978–79): 267–74.

Bagdikian, Ben H. "How Communications May Shape Our Future Environment." In *Mass Media: Forces in Our Society*, ed. Francis Voelker and Ludmila Voelker. New York: Harcourt Brace Jovanovich, 1972.

———. *The Information Machines*. New York: Harper & Row, 1971.

Bell, Daniel. *The Coming of Post-Industrial Society*. New York: Basic Books, 1973.

Bell Telephone. "Harnessing Light for Telecommunications." News release, January 23, 1980.

Bemer, Robert W. "Computers and Our Society." *Jurimetrics Journal* 15, no. 1 (Fall 1974): 43–55.

Benton, John B. "Electronic Funds Transfer: Pitfalls and Payoffs." *Harvard Business Review* 55, no. 4 (July-August 1977): 16–17.

Berkowitz, L., and E. Rawlings. "Effects of Film Violence on Inhibitions against Subsequent Aggression." *Journal of Abnormal and Social Psychology* 66 (1963): 405–12.

Berman, Paul J. "Computer or Communications? Allocation of Functions and the Role of the Federal Communications Commission." In *High and Low Politics: Information Resources for the 80s*. Cambridge, Mass: Ballinger, 1977.

———, and Anthony G. Oettinger. "The Medium and the Telephone: The Politics of Information Resources." In *High and Low Politics: Information Resources for the 80s*. Cambridge, Mass: Ballinger, 1977.

Blumler, J. G., and D. McQuail. *Television in Politics: Its Uses and Influences*. Chicago: University of Chicago Press, 1969.

Bortnick, Jane, ed. "Information and Communications: A Chautauqua for Congress." *Journal of Library Automation* 12, no. 3 (1979): 242–59.

Bowes, John E. "Japan's Approach to an Information Society: A Critical Perspective." Reprinted in *Mass Communication Review Yearbook*, vol 2, ed. Cleaveland G. Wilhoit. Beverly Hills, Calif.: Sage, 1981.

Brancatelli, Joe. "The Home Video Revolution." *EastWest*, supplement, n.d.

Branscomb, Lewis M. "The Electronic Library." *Journal of Communication* 31, no. 1 (Winter 1981): 143–50.

———. "Future Computer: The Shock Will Be Extra Ordinary—And, One Hopes, Benevolent." *Across the Board* 16 (March 1979): 61–68.

———. "Information: The Ultimate Frontier." *Science* 203, no. 4376 (1979): pp. 143–47.

Brewer, Garry D. "On the Theory and Practice of Innovation. " *Technology in Society* 2 (1980): 337–63.

Brzezinski, Zbigniew. *Between Two Ages*. New York: Viking, 1970.

Buck, Elizabeth. "Media Imperialism in Philippine Television." Paper presented at the annual meeting of the International Communication Association, Minneapolis. Minnesota, May 21–25, 1981.

Burnham, David. "U.S. Is Worried by World Efforts to Curtail Flow of Information." *New York Times*, February 26, 1978.

Burns, Scott. "Electronic Mail Will Zap the Postal Zip Code." *The Spokesman-Review*, April 26, 1981, p. B–16.

Busby, Linda J. "Sex-role Research on the Mass Media." *Journal of Communication* 25, no. 4 (Autumn 1975): 107–31.

Bushkin, Arthur A., and Jane H. Yurow. "The Foundation of United States Information Policy." A United States Government Submission to the High-Level Conference on Information, Computer, and Communications Policy. Organization for Economic Cooperation and Development, October 6–8, 1980, Paris, France. National Telecommunications and Information Administration special Publication (NTIA-SP-80-8). Washington, D.C.: Department of Commerce, 1980.

Carson, Pat. "Dirty Tricks: The Confessions of a Cable Franchise Negotiator." *Panorama* 2, no. 5 (May 1981): 56–59, 80.

Cater, Douglass. "The Survival of Human Values." *Journal of Communication* 31. no. 1 (Winter 1981): 190–94.

"CBS Cable; Playing on the Upscale." *Broadcasting*, September 1, 1980.

"CBS Going Live with Teletext in L.A. Text." *Broadcasting*, November 17, 1980, p. 21.

Chaffee, Steven H. and Jack M. McLeod. "Individual vs. Social Predictors of Information Seeking." *Journalism Quarterly* 50, no. 2 (Summer 1973): 237–45.

Chanin, Abe S. "Officials of England's 'Electronic Newspaper'

Predicts No Early Demise of Print Media." *Presstime* 2, no. 12 (December 1980): 6–7.

Cherry, Susan Spaeth. "Telereference: The New TV Information Systems." *American Libraries* 11, no. 2 (Ferbuary 1980); 94–98, 108–10.

Chesebro, James. "A Construct for Assessing Ethics in Communication." *Central States Speech Journal* 20 (Summer 1969): 104–14.

Chisman, Forrest. "Options for Federal Role with Regard to Advanced Telecommunications Systems and Services." Paper presented at the U.S. Congress Office of Technology Assessment Conference on Communications and Rural America, Washington, D.C., November 15–17, 1976.

Chowdhury, Aminur R., and J. Barry DuVall. "Toward a More Humane Technology." *Man/Society/Technology* 37, no. 3 (December 1977): 18–21.

Christians, Clifford G. "Fifty Years of Scholarship in Media Ethics." *Journal of Communication* 27, no. 4 (Autumn 1977): 19–29.

Clarke, P., and F. G. Kline. "Media Effects Reconsidered: Some New Strategies for Communication Research." *Communication Research* 1 (1974): 224–40.

———, and P. Palmgreen. "Media Use, Political Knowledge and Participation in Public Affairs." Paper presented at the meeting of the International Sociological Association, 1974.

Clippinger, John H. *Who Gains by Communications Development? Studies of Information Technologies in Developing Countries.* Working Paper 76–1, Program on Information Technologies and Public Policy. Cambridge, Mass.: Harvard University, 1975.

Cochran, Thomas C. "Media as Business: A Brief History." *Journal of Communication* 25 no. 4 (Autumn 1975): 155–65.

Coldevin, Gary O. "Anik I and Isolation: Television in the Lives of Canadian Eskimos." *Journal of Communication* 27, no. 4 (Autumn 1977): 145–53.

Collins, Hugh. "Forecasting the Use of Innovative Telecommunications Services." *Futures* 12, no. 2 (1980): 106–12.

Collins, W. A., and S. A. Zimmermann. "Convergence and Divergent Social Cues: Effects of Televised Aggression on Children." *Communication Research* 2 (1975): 331–46.

Communications Media Center. *The New World Information Order: Issues in the World Administrative Radio Conference and Transborder Data Flow.* New York: Communications Media Center, New York Law School, 1979.

Compaine, Benjamin M., ed. *Who Owns the Media? Concentration of*

Ownership in the Mass Communications Industry. New York: Harmony Book, 1979.

"Computer Links Stanford Officials in Test of Electronic Mail System." *Chronicle of Higher Education* 11, no. 7 (October 1980): 5.

COMSAT. *Comsat Guide to the Intelsat, Marisat, and Comstar Satellite Systems.* Washington, D.C.: Office of Public Affairs, COMSAT, n.d.

"COMSAT Unveils Its Plans for Satellite-to-Home TV." *The Spokesman-Review*, December 18, 1980, p. 36.

Comstock, George. "The Impact of Television on American Institutions." *Journal of Communication* 28, no. 2 (Spring 1978): 12–28.

Continental Telephone Corporation. "Continental Telephone Regroups, Adds Telecommunications Vice President." Newsrelease, January 7, 1980.

Control of the Direct Broadcast Satellite: Values in Conflict. Palo Alto, Calif.: Aspen Institute for Humanistic Studies, 1974.

Converse, P. E. "Information Flow and the Stability of Partisan Attitudes." *Public Opinion Quarterly* 26 (1962): 578–99.

Cooney, John E., and Carol Hymowitz. "Westinghouse-Teleprompter Link May Solve Problems for Both Firms." *Wall Street Journal*, October 16, 1980, pp. 29, 41.

Council on Learning, ed. *The Communications Revolution and the Education of Americans.* New Rochelle, N.Y.: Change Magazine Press, 1980.

Crock, Stan. "TV Fans' Latest Toy Is a Personal Satellite 'Dish' That Snares Signals Sent Up by Cable Networks." *Wall Street Journal*, July 10, 1980.

Culbertson, Hugh M. "Words vs. Pictures: Perceived Impact and Connotative Meaning." *Journalism Quarterly* 51, no. 2 (Summer 1974): 266–37.

Danowski, James A. "Adult Life-Course Socialization to Communication Technologies: Aging, Cohort and Period Effects." Paper presented to the Mass Communication Division of the International Communication Association annual meetings, Minneapolis, Minnesota, May 1981.

Dator, James A. "Identity, Culture and Communication Futures." *Futurics* 3, no. 3 (1979): 207–24.

Dicks, Dennis. "From Dog Sled to Dial Phone: A Cultural Gap?" *Journal of Communication* 27, no. 4 (Autumn 1977): 120–29.

"Direct Satellite B'cast Service Is On Sked: COMSAT." *Variety* 301, no. 2 (November 12, 1980): 55.

Dobbert, Marion Lundy. "Proposals for an Anthropologically Based Pan-Cultural Ethical Process." *Futurics* 3, no. 2 (1979): 131–43.

Dominick, J. R., and G. E. Rauch. "The Image of Women in Network TV Commercials." *Journal of Broadcasting* 16 (1972): 259–65.

Dommermuth, W. P. "How Does the Medium Affect the Message?" *Journalism Quarterly* 51 (1974): 441–47.

Donohew, Lewis; Leonard Tipton; and Roger Haney. "Analysis of Information-Seeking Strategies." *Mass Communication Review Yearbook 1980*, pp. 383–89.

Donohue, G. A.; P. J. Tichenor; and C. N. Olien. "Gatekeeping: Mass Media Systems and Information Control." In *Current Perspectives in Mass Communication Research*, ed. F. G. Kline and P. J. Tichenor. Beverly Hills, Calif.: Sage, 1974, pp. 41–69.

———. "Mass Media Functions, Knowledge and Social Control." *Journalism Quarterly* 50, no. 4 (Winter 1973): 652–59.

Ducey, Rick. "Information as a Public Good: The Social Resource of Broadband Communication Networks." Presented to the International Communication Association, Mass Communication Division, Minneapolis, Minnesota, May 21–25, 1981.

Edelstein, Alex S. "Information Societies: An Introduction." Paper presented at the Japan-U.S.A. Conference on "Comparing Information Societies, Directions for Research" at the University of Washington, 1980.

———; John E. Bowes; and Sheldon M. Harsel. *Information Societies: Comparing the Japanese and American Experience.* Seattle: University of Washington Press, 1978.

Edward, Kenneth. "The Electronic Newspaper." *The Futurist* 12, no. 2 (1978): 79–84.

———. "Information Without Limit Electronically." In *Readings in Mass Communication: Concepts and Issues in the Mass Media*, ed. Michael Emery and Ted Curtis Smythe. Dubuque, Iowa: Wm. C. Brown, 1980, pp. 202–16.

Eimbinder, Jerry and Eric. "Electronic Games: Space-Age Leisure Activity." Part 2. *Popular Electronics* 18, no. 5 (November 1980): 89–93.

Elliott, David, and Ruth Elliott. *The Control of Technology.* London: Wykeham Publications, 1976.

Ellul, Jacques. *The Technological Society.* Trans. John Wilkinson. New York: Vintage Books, 1967.

Enberg, Ole. "Who Will Lead the Way to the 'Information Society'?" *Impact of Science on Society* 28, no. 3 (1978): 283–96.

Evans, Laurence. *The Communication Gap: The Ethics and Machinery of Public Relations and Information*. London: Charles Knight, 1973.

Executone Inc. (Member of Continental Telephone System). "Announcing the Intelligent Intercom." 1979; "At Last, an Intelligent Telephone That Gives Everyone The Same Advantages as the Big Operators." 1980; "The Executone D-1000 Telephone." 1977; "Introducing a Big Phone System in a Little Package." 1979; "Own Your Own Phone Company." 1977; "You Can Have the Nurse Call System of Tomorrow Today." 1979; "Your Employees are Talking You Out of Thousands of Dollars a Year." 1979; (All publicity brochures.)

"Eyes on Texas as AT&T Plans Test in Austin." *Presstime* 2, no. 12 (December 1980): 17.

Fairchild Industries. "Continental Telephone, Fairchild Industries Partnership Plans Expansion of American Satellite System." Fairchild news release, July 31, 1980.

"The Fancy New Video Games." *Changing Times* 32, no. 11 (November 1978): 43–44.

Fay, Timothy. "Toward Development of Public Telecommunications Centers." Proc. of the Association for Education Data Systems' 16th Annual Convention, May 16–19, 1978, Atlanta, Georgia, pp. 32–39.

Fedler, Fred. "The Media and Minority Groups: A Study of Adequacy of Access." *Journalism Quarterly* 50, no. 1 (Spring 1973): 109–117.

Ferkiss, Victor C. *Technological Man: The Myth and the Reality*. New York: George Braziller, 1969.

Ferrell, Tom. "Libraries Face Up to the New Technological Imperatives." *New York Times*, June 29, 1980, sec. E, p. 20.

Fishman, William L. "International Data Flow: Personal Privacy and Some Other Matters." Paper presented before the Fourth International Conference on Computer Communication, September 26–29, 1978, Kyoto, Japan.

Fitzpatrick, Paul. "Looking Ahead to the Eighties: The Decade's Top Three Events." *CableVision*, December 17, 1979, p. 208.

Foldy, Reginald. "Collective Depression, New Malady of an Information-saturated Society." *Impact of Science on Society*, 30, no. 2 (April-June 1980): 87–92.

Foote, Dennis R. "Satellite Communication for Rural Health Care in Alaska." *Journal of Communication* 27, no. 4 (Autumn 1977): 173–82.

Frank, Ronald A. "TI Ties into IBM Terminals." *Datamation* 27, no. 4 (April 1981): 71, 90.

———. "Telecom Crisis Planning." *Datamation* 27, no. 4 (April 1981): 90.

Gaines, B. R. "Man-Computer Communication—What Next?" *International Journal of Man-Machine Studies* 10, no. 3 (1978): 225–32

Gammill, Robert C. "Personal Computers for Science in the 1980's." Paper presented at the Conference on Problems of the 80's: The Oregon Report on Computing, March 21, 1978, Portland, Oregon.

Gerbner, G. "Comments on 'Measuring Violence on Television: The Gerbner Index'," by Bruce M. Owen. Staff Research Paper, Office of Telecommunications Policy. Manuscript, 1972. Annenberg School of Communications, University of Pennsylvania.

———, and L. Gross. "Living with Television: The Violence Profile." *Journal of Communication* 26, no. 2 (1976): 173–99.

———; M. F. Eleey; M. Jackson-Beeck; S. Jeffries-Fox; and N. Signorielli. *Violence Profile No. 8: Trends in Network Television Drama and Viewer Conceptions of Social Reality.* Philadelphia: Annenberg School of Communications, 1977.

———; and William H. Melody. *Communications Technology and Social Policy: Understanding the New "Cultural Revolution."* New York: John Wiley & Sons, 1973.

Goldhammer, Herbert. "The Social Effects of Communication Technology." Rand Corporation Paper, 1970.

Golding, Peter, and Graham Murdock. "Theories of Communication and Theories of Society." *Mass Communication Review Yearbook 1980*, pp. 59–76.

Goldman, Michael. "New Technology: Destructive, Or Not?" *Variety* 301, no. 11 (January 14, 1981): 15, 122.

Goldman, Ronald J. "Demand for Telecommunication Services in the Home." Paper presented at the Annual Meeting of the International Communication Association, May 1–5, 1979, Philadelphia, Pennsylvania.

Goldmark, Peter C. "Tomorrow We Will Communicate to Our Jobs." In *Human Connection and the New Media*, ed. Barry N. Schwartz. Englewood Cliffs, N.J.: Prentice-Hall, 1973, pp. 172–79.

Goldsmith, Tom. "COMSAT Bird Bid Opens Satellite-to-Home Era." *Variety* 301, no. 8 (December 24, 1980): 40.

Goulet, Denis. "Technology and Human Ends: What Are the Choices? *Social Education,* 1979, pp. 427–32.

Gruenberger, Fred. "Making Friends with User-Friendly." *Datamation*, January 1981 , pp. 108–10.

Haberer, Joseph. "Technology and the Emerging Future: A Framework for Normative Theory." *The Human Context* 7, no. 1 (Spring 1975): 130–35.

Halpern, Werner I. "Turned-on Toddlers." *Journal of Communication* 25, no. 4 (Autumn 1975): 66–70.

Hanauer, Gary. "Conventional Wisdom." *Passages*, May 1981, pp. 7–8.

Harms, L. W. "The Right to Communicate and Its Implementation Within a New World Communication Order." Paper presented at the UNESCO Experts Meeting on the Right to Communicate, Stockholm, Sweden, May 8–12, 1978.

Harsel, Sheldon, "Communication Research in Information Societies: A Comparative View of Japan and the United States." Reprinted in *Mass Communications Review Yearbook*, vol 2.

Hauser, Gustave M. "The Wave of the Future: Teleshopping." No source given; reprint from Warner Amex Co., 24–27

Hayashi, Yujiro. *Johoka Shakai* (Informational Societies). Tokyo: Kodansha Gendai Shinsho, 1969.

Heller, Caroline. "The Resistible Rise of Video: Some Thoughts on a Technology and Social Change." *Educational Broadcasting International* 11, no. 3 (September 1978): 133–35.

Helmer, Olaf. *Social Technology*. New York: Basic Books, 1966.

Hetman, F. *Society and Assessment of Technology*. OECD, 1973.

Heun, Dick and Linda, and Vi Martin. "Computer Modeling of Comprehension in Listening and Reading." Paper presented at the annual meeting of the International Communication Association, Minneapolis, Minnesota, May 1981.

Hill, G. Christian, and Brenton R. Schlender. "Times Mirror to Buy Denver Post: $95 Million Price Surprises Experts." *Wall Street Journal*, October 23, 1980, p. 29.

Hiltz, Starr Roxanne, and Murray Turoff. *The Network Nation: Human Communication via Computer*. Readings, Mass.: Addison-Wesley, 1978.

Hinshaw, Mark L. "Wiring Megalopolis: Two Scenarios." In *Communications Technology and Social Policy: Understanding the New "Cultural Revolution,"* ed. George Gerbner, Larry P. Gross, and William H. Melody. New York: John Wiley & Sons, 1973, pp. 305–17.

Holmes, Edith. "Transborder Data Taking Back Seat." *Computerworld*, October 24, 1977.

Homet, Roland S., Jr. *Politics, Cultures, and Communication*. New York: Praeger, 1979.

Hornik, Robert C. "Mass Media Use and the 'Revolution of Rising

Frustrations': A Reconsideration of the Theory." *Communication Research* 4, no. 4 (1977): 387–413.

"How Could $21.3 Million Disappear from the Nation's 11th Largest Commercial Bank?" *Los Angeles Times*, February 7, 1981.

Hughes, Michael. "The Fruits of Cultivation Analysis: A Reexamination of Some Effects of Television Watching." *Public Opinion Quarterly*, 1980, pp. 401–415.

Hull, Joseph A. "Fiber Optic Communications Technology: A Status Report." Paper presented at the U.S. Congress Office of Technology Assessment Conference on Communications and Rural America, Washington, D.C., November 15–17, 1976.

Hunt, Dave. "A Definitional Model of Information Transfer as a Communication Process." Paper presented at the International Communication Association, 31st Annual Conference, Minneapolis, Minnesota, May 21–25, 1981.

Hurwitz, Sol. "On the Road to Wired City." *Harvard Magazine*, September-October 1979, pp. 18–23.

Huskell, Andrew. "The World of the Word." Paper presented at the Public Relations World Congress, Boston, Massachusetts, August 1976.

Hyman, Barry. "Technology Policy Analysis: Bridging the Technology Gap." Washington Public Policy Notes. Institutes of Governmental Research and the University of Washington, n.d.

Ingalls, Richard E. "Intelligent Video Disc as a Major Component of Individualized Instruction." 1977.

Irwin, M. R., *The Telecommunications Industry: Integration vs. Competition*. New York: Praeger, 1971.

Ito, Youichi. "The *Johoka Shakai* Approach to the Study of Communication in Japan." *Keio Communication Review*, March 1980, pp. 13–40.

Japan Computer Usage Development Institute Computerization Committee (1972). "The Plan for an Information Society: A National Goal Toward the Year 1000." Final Report. Tokyo. Reprint in *Data Exchange*, July-August 1973.

Johansen, Robert, and Robert DeGrasse. "Computer-Based Teleconferencing: Effects on Working Patterns." *Journal of Communication* 29, no. 3 (1979): 30–41.

Journal of Communication 31, no. 1 (Winter 1981). Devoted to the information society.

Kahn, Herman, and B. Bruce-Briggs. *Things to Come: Thinking about the Seventies and Eighties*. New York: Macmillan, 1972.

Kalmen, Harry Jr. "The Problems of Privacy in the Year 2000." *Daedalus* 96, no. 3 (Summer 1967): 876–82.

Katzman, N. "The Impact of Communication Technology: Promise and Prospects." *Journal of Communication* 24, no. 4 (1974): 47–58.

Keshishoglou, John E. "Cable Television: Friend or Enemy of the Future?" *EBU Review* 27 (September 1976): 18–20.

Kimbel, Dieter. "An Assessment of the Computer-Telecommunications Complex in Europe, Japan, and North America." In *Communications Technology and Social Policy: Understanding the New "Cultural Revolution,"* ed. George Gerbner, Larry P. Gross, and William H. Melody. New York: John Wiley & Sons, 1973, pp. 147–164.

Kirby, M. D. "Eight Years to 1984: Privacy and Law Reform." *Rutgers Journal of Computers and the Law* 5, no. 2 (1976): 487–502.

Kirk, Joan. "The Big Picture." *Passages*, May 1981, pp. 11–16.

Kleinjans, Everett. "Communication and Change in Developing Countries." Paper No. 12. Hawaii University, Honolulu, EastWest Center, July 1975. ERIC ED 163 534.

Knight-Ridder Newspapers, Inc. "Viewtron." Newsrelease, April 17, 1979.

Knoll, Steve. "An RCA Primer on Homevid's Legal Web of Program Rights." *Variety* 301, no. 2 (November 12, 1980): 48.

Koo, Charles M. "Mass Media Decision in China's Post Mao Zedong Modernization Program: Some Unanticipated Consequences." Paper presented at the 31st International Communication Association International Convention, Intercultural Communication Division, Minneapolis, Minnesota, May 21–25, 1981.

Koyama, Kenichi. "Joho Shakai Ron Josetsu" (Introduction to Information Society Theory). *Chuo Koron Keiei Mondai*, Winter 1968, pp. 80–105.

LaBlanc, Robert. "The Future of Communications: The Inivitation of the Eighties and Nineties." Paper presented before the Illinois Telephone Association, June 23, 1980, Lincolnshire, Illinois.

———. "Telecommunications in a Decade of Scarcity." Paper presented before the Fixed Income Analysts Society, Inc., May 14, 1980, New York, New York.

"Latest Videodisk Entrant." *Broadcasting*, June 16, 1980, p. 68.

LeDuc, Don R. "West European Broadcasting Policy: Implications of New Technology." *Journal of Broadcasting* 23, no. 2 (Spring 1979): 237–44.

Lee, John A. R. *Towards Realistic Communication Policies: Recent Trends and Ideas Compiled and Analysed.* Reports and Papers on Mass Communication, No. 76. Place de Fontenoy, Paris: Department of Mass Communication, UNESCO, 1976.

Levy, Mark R. "Home Video Recorders: A User Survey." *Journal of Communication* 30, no. 4 (Autumn 1980): 23–27.

Lewis, Elaine. "Visual Communication as a Human Information Processing System: Toward Definition and Evaluation." Paper presented at the International Communication Association Conference, Information Systems Division, May 1981.

Logue, Timothy J. "Teletext: Towards an Information Utility?" *Journal of Communication* 29, no. 4 (Autumn 1979): 58–65.

Lowenhaupt, Thomas J. "Integrating Transportation Information Systems with Cable Communications." Transcript of presentation before the 29th Annual convention of the National Cable Television Association, May 1980, pp. 60–63.

Magarrell, Jack. "Five Universities Protest U.S. Efforts to Limit International Exchange of Research Data." *Chronicle of Higher Education* 22, no. 10 April 27, 1981): 1, 8.

Maguire, W. Terry, and Douglas R. Watts. "Newspaper Use of Cable, Telephone and Satellites: Some Legal and Related Business Considerations." *Presstime* 2, no. 12 (December 1980): A1-A7.

Manitoba Telephone System. *Exploring the Wired City*. Manitoba, Canada: Manitoba Telephone System, n.d.

Manning, Robert. "Data Is Wealth and Power." Review of *The Geopolitics of Information* by Anthony Smith. *New York Times Book Review*, December 7, 1980, pp. 15, 38.

Marker, Gerald W. "Exploring the Relationship between Technology and Social Values." Paper presented at NSF Workshops on New Developments in Science and Social Science, April 1978, Ann Arbor, Michigan.

Martin, James. *Future Developments in Telecommunications*. Englewood Cliffs, N.J.: Prentice-Hall, 1977.

———. *Telecommunications and the Computer*. Englewood Cliffs, N.J.: Prentice-Hall, 1969.

———, and Adrian R. D. Norman. *The Computerized Society*. Englewood Cliffs, N.J.: Prentice-Hall, 1970.

Masmoudi, Mustapha. "The New World Information Order." *Journal of Communication* 29 no. 2 (Spring 1979): 172–85.

Matleiart, Armand. "Modern Communication Technologies and New Facets of Cultural Imperialism." *Instant Research on Peace and Violence* 1 (1973): 9–27.

McCarthy, Elizabeth; Thomas S. Langner; Joanne C. Gersten; Jeanne G. Eisenberg; and Lida Orzeck. "Violence and Behavior Disorders." *Journal of Communication* 25, no. 4 (Autumn 1975): 71–85.

McKay, K. G. "The Network." *Science and Technology*, 1968.

McLeod, J. M.; S. H. Chaffee; and H. S. Eswara. "Family Communication Patterns and Communication Research." Paper presented at the meeting of the Association for Education in Journalism, Iowa City, Iowa, August 1966.

McLeod, J. M.; S. Ward; and K. Tancill. "Alienation and the Uses of the Mass Media." *Public Opinion Quarterly* 29 (1965): 585–94.

McLuhan, Marshall, and Bruce Powers. "Electronic Banking and the Death of Privacy." *Journal of Communication* 31, no. 1 (Winter 1981): pp. 164–69.

McNeil, J. C. "Feminism, Femininity, and the Television Series: A Content Analysis." *Journal of Broadcasting* 19, no. 3 (1975): 259–69.

McNeil, Jean C. "Whose Values?" *Journal of Broadcasting* 19, no. 3 (Summer 1975): 295–96.

Messaris, Paul. "Family Processes and the Social Functions of Television." Paper presented at the Temple University Conference on Culture and Communication, Philadelphia, April 9, 1981.

Mesthene, Emmanuel G. *Technological Change: Its Impact on Man and Society*. New York: Mentor, 1970.

Ministry of Posts and Telecommunications, Communications Policy Division. "Joho Ryutsyu no Keiryo" (The Methods for the Measurement of Information Flow). 1975.

———. "Information Flow Census in Japan—A Quantitative Study of Information Societies." Unpublished English paper distributed at an OECD conference, 1978.

Morris, DuBois S. *Information Technology: Initiatives for Today—Decisions That Cannot Wait. Some Major Problem Areas and Leadership Options*. New York, N.Y.: Conference Board, 1972.

Mumford, Lewis. *The Myth of the Machine: Technics and Human Development*. New York: Harcourt, Brace and World, 1966.

"NASA Pushing for Next Generation of Satellites." *Broadcasting*, June 16, 1980, p. 68.

National Academy of Sciences. *Technology: Processes of Assessment and Choice*. Washington, D. C.: Government Printing Office, 1969.

Nelso, Richard A. "The New Information and Programming Technologies: What They Mean for Audiences." Paper presented at the Popular Culture Association and American Culture Association, Louisville, Kentucky, April, 1982.

Network Analysis Corporation. *Activities in Computers and Communications*. Great Neck, N.Y.: Network Analysis Corporation, n.d.

Newman, Joseph, dir. ed. *Wiring the World: The Explosion in Communications.* Washington, D.C.: U.S. News & World Report, 1971, p. 17.

"Nickelodeon: Warner's Children's Channel." *TVC,* April 1, 1979. Reprint.

"Nielsen Looks at Cable Viewers: Cable Homes Don't Watch More, Paycablers Do; Indies Are Hurt." *Variety,* 301, no. 2 (November 12, 1980): 2.

Nilsen, Svein Erik. "The Use of Computer Technology in Some Developing Countries." *International Social Science Journal* 31, no. 3 (1979): 513–28.

Norman, Colin. *Hard Choices Worldwatch Paper 21.* Washington, D.C.: Worldwatch Institute, 1978.

"Not Everyone Pays for Pay-TV." *TV Guide* 29, no. 10 (March 7, 1981): 7–8, 10–11.

O'Brien, Rita Cruise. "Specialized Information and Global Interdependence Problems of Concentration and Access." Paper presented at the Annual Meeting of the International Institute of Communications, September 9–13, 1979, London, England.

O'Connor, John L. "Would Viewers Be Willing to Pay for Culture?" *New York Times,* June 8, 1980, sec. 2, pp. 1, 31.

Oden, Teresa, and Christine Thompson, eds. *Computers and Public Policy. Proceedings of the Symposium Man and the Computer.* Hanover, N.H.: Dartmouth College, 1977.

Oettinger, Anthony G. *Elements of Information Resources Policy: Library and Other Information Services.* Revised edition. Program on Information Resources Policy, Center for Information Policy Research. Cambridge Mass.: Harvard University Press, 1975.

———. "Information Resources: Knowledge and Power in the 21st Century." *Science* 209, no. 4 (1980).

———; Paul J. Berman; and William H. Read. *High and Low Politics: Information Resources for the 80s.* Cambridge, Mass.: Ballinger, 1977.

———, and Peter D. Shapiro. *A Dialogue on Technology Assessment: The Video Telephone Critique and Rejoinder.* Publication No. 75-1, Program on Information Technologies and Public Policy. Cambridge, Mass.: Harvard University Press, 1975.

Olson, David R. "Oral and Written Language and the Cognitive Processes of Children." *Journal of Communication* 27, no. 3 (Summer 1977): 10–36.

Orwell, George. *Nineteen Eighty-Four.* New York: Harcourt, 1949.

Parker, Edwin B. *Procedural Conference on Computer/Telecom-*

munications Policy. Paris: Organization for Economic Cooperation and Development, 1975.

———. "Technology Assessment or Institutional Change?" In *Communications Technology and Social Policy: Understanding the New "Cultural Revolution,"* ed. George Gerbner, Larry P. Gross, and William H. Melody. New York: John Wiley & Sons, 1973, pp. 533–46.

———, and Donald A. Dunn. "Information Technology: Its Social Potential." *Science* 176, no. 4042 (January 30, 1972): 1392–99.

———, and Marc Porat. "Social Implications of Computer/Telecommunications Systems." Paper presented at the Conference on Computer/Telecommunications Policies, February 1975, Paris, France. ERIC ED 102 978.

Pelton, Joseph N. "The Future of Telecommunications." *Journal of Communication* 31, no. 1 (Winter 1981): 177–89.

———. *Global Talk.* Netherlands: Sijthoff & Noordhoff, 1981.

———. "INTELSAT Initiatives for the 1980's and Their Implications for the Third World." Mimeograph.

Pierce, John R. "Communications Technology and the Future." *Daedalus* 94, no. 2 (Spring 1965): 506–17.

"Pirate-Free Bird Keys Comsat Bid for DBS Success." *Variety* 301, no. 8 (December 24, 1980): 40.

Pool, Ithiel de Sola. "International Aspects of Computer Communications." *Telecommunications Policy,* December 1976, 33–51.

———, ed. *The Social Impact of the Telephone.* Cambridge, Mass.: MIT Press, 1977.

Porat, Marc U. "Communication Policy in an Information Society." In *Communications for Tomorrow: Policy Perspectives for the 1980s,* ed. Glen O. Robinson. New York: Praeger, 1978, pp. 3–60.

———. "Global Implications of the Information Society." *Journal of Communication* 28, no. 1 (Winter 1976): 70–80.

Poulos, Rita Wicks; Eli A. Rubinstein; and Robert M. Liebert. "Positive Social Learning." *Journal of Communication* 25, no. 4 (Autumn 1975): 90–97.

Price, Charlton R., and Elaine B. Kerr. "Electronic 'Connectedness': Its Meaning for Personal and Social Disabilities." *Bulletin of the American Society for Information Science* 4, no. 5 (January 1978): 19–20.

Public Service Satellite Consortium. "Commonly Used Terms in Telecommunications." Mimeograph.

"Qube's Two-way Facility: New Dimensions for Books." *Columbus Citizen-Journal,* May 19, 1980, p. 15.

Reinhardt, John E. "Global Communications in the Nineteen Eighties: An American Perspective." Keynote address at the International Conference on World Communications, May 12, 1980, Annenberg School of Communications, Philadelphia, Pennsylvania.

Rensselaer Polytechnic Institute. *Impact of Future Cable Television Technology*. Final Summary Report. Troy, N.Y.: Rensselaer Polytechnic Institute, 1976.

Report on TAMA CCIS Experiment Project in Japan: 1978. Tokyo, Japan: [Visual Information System Development Association], 1979.

Research Institute of Telecommunications and Economics (1968). *Sangyoka Igo no Shakai ni Okeru Joho to Tsushin* (Information and Communication in a Postindustrial Society).

"Researching Corporate Videoconferencing." *Record*, Winter 1980, pp. 1, 3–4.

Rice, Ronald E. "Computer Conferencing." In *Progress in Communication Sciences*, vol. 2, ed. B. Dervin and M. Voigt. New Jersey: Ablex, 1980.

————. "The Impacts of Computer-Mediated Organizational and Interpersonnel Communication." In *Annual Review of Information Science and Technology*, vol. 15, ed. M. Williams. White Plains, N.Y.: Knowledge Industry Pubs., 1980.

————, and Edwin B. Parker. "Telecommunications Alternatives for Developing Countries." *Journal of Communication* 29, no. 4 (Autumn 1979): 125–36.

————, and Everett M. Rogers. "Facilitation of Computer-Mediated Communication: Innovation in the Organization." Paper presented to ICA, Minneapolis, Minnesota 1981.

Robinson, Glen O., ed. *Communications for Tomorrow: Policy Perspectives for the 1980's*. New York: Praeger, 1978.

Rogers, Everett M. *Modernization Among Peasants: The Impact of Communication*. New York: Holt, Rinehart & Winston, 1969.

Rokeach, Milton. "Change and Stability in American Value Systems: 1968–1971." *Public Opinion Quarterly* 38, no. 2 (1974): 222–38.

————. *The Nature of Human Values*. New York: Free Press, 1973.

————. "Value Theory and Communication Research: Review and Commentary." In *Communication Yearbook III*, ed. Dan Nimmo. Brunswick, N.J.: Transaction Books, 1979, pp. 7–28.

Ryu, Jung S. "The Perception of Shogun in a Midwestern Community." Paper presented at the annual meeting of the International Communication Association, Minneapolis, Minnesota, May 1981.

Salvaggio, Jerry L. "The Impact of Videotex on the Newspaper Industry" Paper presented at regional meeting of Association for Education in Journalism, Atlanta, Georgia. February 1982.

────── and Susan K. Trettevik. "Transition to the Wired World: A Model for the Study of Information Inequity in an Information Society." Paper presented at the International Communication Association meetings, Minneapolis, Minnesota, May 1981.

──────, "An Assessment of Japan as an Information Society in the 1980's" in *Communications and the Future*, Ed. Howard F. Didsbury, *World Future* Society, Bethesda. MD: 1982.

Sanders, Ray W. "Toward Total Deregulation." *Datamation* 27, no. 4 (April 1981): 273–74, 276.

Sargent, L. W., and G. H. Stempel. "Poverty, Alienation and Mass Media Use." *Journalism Quarterly* 45 (1968): 324–26.

Schadewald, Robert J. "The Data Game: Business and the Micro Computer." *Passages*, May 1981, pp. 19–34.

Schatz, Willie. "Much Ado About Nothing." *Datamation* 27, no. 4 (April 1981): 48.

Schiller, Herbert I. "Communication Accompanies Capital Flows." Paper prepared for the International Commission for the Study of Communication Problems, UNESCO. Document No. 47.

──────. "Computer Systems: Power for Whom and for What?" *Journal of Communications* 28, no. 4 (1978): 184–93.

──────. "Decolonization of Information: Efforts Toward a New International Order." *Latin American Perspectives* 5, no. 1 (Winter 1978): 35–48.

──────. "Media and Imperialism." *Revue Francaise D'Etudes Americanes* 6 (October 1978): 269–81.

──────. *Who knows: Information in the Age of the Fortune 500*. Norwood, N.J.: Ablex, 1981.

Schlafly, Hubert J. "A Review of the Unwired Nation." Paper presented at the Telecommunications Policy Research Conference, April 21–24, 1976, Airlie, Virginia. ERIC ED 122 771.

Schmedel, Scott R. "TV Systems Enabling Viewers to Call Up Printed Data Catch Eye of Media Firms." *Wall Street Journal*, July 24, 1979, p. 46.

Schramm, Wilbur. *Communication Satellites for Education, Science and Culture*. Reports and Papers on Mass Communication, No. 53. Place de Fontenoy, Paris: Department of Mass Communication, UNESCO, 1967.

──────. "Some Possible Social Effects of Space Communication." In *Communications in the Space Age: The Use of Satellites by the*

Mass Media. Amsterdam, Netherlands: UNESCO, 1968, pp. 11–29.

Scott, Robin. "New Media—New Messages?" *EBU Review* 30, no. 3 (1979): 61–66.

Settle, Tom. "The Moral Dimension in Political Assessments of the Social Impact of Technology." *Philosophy of the Social Sciences* 6, no. 4 (December 1976): 315–34.

Shapiro, Neil. "Games on Computers." *Popular Mechanics* 152, no. 2 (August 1979): 69.

Sharp, Elaine. "Citizen Perceptions of Channels for Urban Service Advocacy." *Public Opinion Quarterly*, 1980, pp. 363–76.

Shepherd, William G., Jr. "Tayloring Television for a New Audience." *Mainliner*, May 1981, pp. 131, 152–54, 156–58.

Sherman, Charles E., and John Ruby. "Eurovision News Exchange." *Journalism Quarterly*, 51, no. 3 (Autumn 1974): 478–85.

Short, John; Ederyn Williams; and Bruce Christie. *The Social Psychology of Telecommunications*. New York: John Wiley & Sons, 1976.

Silberman, Alphons. "Communication Systems and Future Behavior Patterns." *International Social Science Journal* 29, no. 2 (1977): 337–41.

Smith, Anthony. *The Geopolitics of Information*. New York: Oxford University Press, 1980.

Smith, Ralph Lee. *The Wired Nation. Cable TV: The Electronic Communications Highway*. New York: Harper & Row, 1972.

Smith, Robert Frederick. "A Funny Thing Is Happening to the Library on Its Way to the Future." The Futurist 12 no. 2 (1978): 85–91.

Sofranko, Andrew J.; Frederick C. Fliegel; and Navin C. Sharma. "A Comparative Analysis of the Social Impacts of a Technological Delivery System." *Human Organization* 36, no. 2 (Summer 1977): 193–97.

Spencer, J. R. *Holographic Information Storage and Retrieval*. Final Report.Hatfield Polytechnic (England), 1975.

Spinard, Norman. "Elections: Will the Future Bring Electronic Democracy?" *Seattle Times*, June 29, 1980, sec. A, p. 3.

Sprafkin, Joyce N.; L. Theresa Silverman; and Eli A. Rubinstein. "Reactions to Sex on Television: An Exploratory Study." *Public Opinion Quarterly*, 1980, pp. 303–15.

Stace, W. T. "Values in General." In Stace, *Man Against Darkness*. Pittsburgh: University of Pittsburgh Press, 1967, pp. 67–85.

Steinfield, Charles W. "Explaining Managers' Use of Communication Channels for Performance Monitoring: Implications for New

Office Systems." Paper presented at the Annual Convention of the International Communication Association, Minneapolis, Minnesota, May 21–25, 1981.

Sterling, C. K., and T. R. Haight. *The Mass Media: Aspen Institute Guide to Communication Industry Trends.* New York: Praeger, 1978.

Sutherland, N.S. "The Electronic Oracle." *Times Literary Supplement,* July 30, 1976.

Szuprowicz, Bohdan O. "The World's Top 50 Computer Import Markets." *Datamation,* January 1981, pp. 141–142, 144.

Tan, Alexis S. "A Role Theory: A Dissonance Analysis of Message Content Preferences." *Journalism Quarterly* 50, no. 2 (Summer 1973): 278–84.

Tannenbaum, Jeffry A., and William M. Bulkeley. "Chasing Paper: Device Makers Dream of Electronic Offices, But Obstacles Remain." *Wall Street Journal* 61, no. 105 (March 13, 1981): 1, 13.

Teich, Albert H., ed. *Technology and Man's Future.* 2nd ed. New York: St. Martin's Press, 1977.

Thayer, Lee, ed. *Ethics, Morality and the Media: Reflections on American Culture.* New York: Hastings House, 1980.

Tichenor, P. J.; G. A. Donohue; and C. N. Olien. "Mass Media Flow and Differential Growth in Knowledge." *Public Opinion Quarterly* 34, no. 2 (Summer 1970): 159–70.

Tichenor, Phillip J., George A Donohue; and Clarice N. Olien. "Mass Communication Research: Evolution of a Structural Model." *Journalism Quarterly* 50, no. 3 (Autumn 1973): 419–25.

Toffler, Alvin. *The Third Wave.* New York: William Morrow, 1980.

Tomita, Tetsuro. "Japan: The Search for a Personal Information Medium." *Intermedia* 7, no. 3 (May 1979): 36–38.

Townley, Rod. "It's Getting Scary Up There!" *TV Guide,* April 25, 1981, pp. 38–39.

Trachtenberg, Alan. "Technology and Human Values." *Technology and Culture* 5, no. 3 (1964): 359–76.

Tsuneki, Teruo. "Johoryo Shokutei ni Kansuru Ichi Kosatsu" (Measurement of the Amount of Information). *RITE Review* 4 (1980): 47–67.

Tunstall, Jeremy. *The Media Are American: Anglo-American Media in the World.* New York: Columbia University Press, 1977.

"TV: A Growth Industry Again." *Business Week,* no. 2676 (February 23, 1981): 88–91.

Umecao, Tadao, "Joho Sangyo Ron" (On Information Industries). *Chuo Koron,* March 1963, pp. 46–58.

UNESCO. Advisory Group of Experts in Informations (AGI). Final Report. Paris: UNESCO, September 17–19, 1979.

———. *A Guide to Satellite Communication.* Reports and Papers on Mass Communication, No. 66. Place de Fontenoy, Paris: Department of Mass Communication, UNESCO, 1972.

———. *Informatics: A Vital Factor in Development.* Paris: UNESCO, 1980.

Vail, Hollis. "The Automated Office." *The Futurist* 12, no. 2 (1978): 73–77.

"Video Shack Sextape Sales Dip, But Porno Still the Top Seller." *Variety* 301, no. 2 (November 12, 1980): 48.

Visual Information System Development Association. Hi-Ovis Project: Hardware/Software Experiments, July 1978–March 1979. Interim Report. Tokyo, Japan: Visual Information System Development Association (VISDA)-MITI Juridical Foundation, 1979.

———. Optical Visual Information System: HI-OVIS. Tokyo, Japan: Visual Information System Development Association (VISDA)-MITI Juridical Foundation, 1979.

Ware, Willis H. "Computer Technology: For Better or Worse?" Paper presented at the National Bureau of Standards Conference on Trends in Applications 1977: Computer Security and Integrity, May 19, 1977. Washington, D.C.

———. "Computers and Personal Privacy." Santa Monica, Calif.: Rand Corporation, July 1977.

———. "Computers and Society—The Technological Setting." *Jurimetrics Journal* 14, no. 3 (Spring 1974): 141–57.

Warren, Samuel D., and Louis D. Brandeis. "The Right to Privacy." *Harvard Law Review* 4, no. 5 (December 15, 1980): 124–220.

Waters, Harry F.; Cynthia H. Wilson; and Pete Davies. "TV Turns to Print." *Newsweek,* (July 30, 1979, pp. 73–75.

Weintraub, B. "The Whole Idea of TV May Change." *Washington Star,* December 12, 1979, p. C-2.

"Westar III: Is It Changing into a Cable Bird?" *Broadcasting,* June 16, 1980, p. 66.

Western Electric. "Western's New Atlanta Facility Brings Lightwave System to Life." Newsrelease, September 15, 1980.

Westin, Alan. *Information Technology in a Democracy.* Cambridge, Mass.: Harvard University Press, 1971.

Widmer, Kingsley. "Sensibility Under Technology: Reflections on the Culture of Processed Communications." In *Human Connection and the New Media,* ed. Barry N. Schwartz. Englewood Cliffs, N.J.: Prentice-Hall, 1973, pp. 28–41.

Wiio, Osmo A. "What Is Information? A Conceptual Analysis of Some Basic Words." Paper prepared for the International Communication Association meeting, Information Systems Division session, Minneapolis, Minnesota, May 1981.

Wiley, Richard E. "Family Viewing: A Balancing of Interests." *Journal of Communication* 27, no. 2 (Spring 1977): 188–92.

Williams, Frederick; Herbert S. Dordick; and Frederick Horstmann. "Where Citizens Go for Information." *Journal of Communication* 27, no. 1 (Winter 1977): 95–100.

Williams, Martha, and Ted Brandharat. "Data About Data Bases." *Bulletin of the American Society for Information Science* 3, no. 2 (December 1976).

Williamson, Jeff. "QUBE: Future Tense of Tube?" *Madison Avenue Magazine*, reprint, n.d., pp. 92–96.

Wilson, C. Edward. "The Effect of Medium on Loss of Information." *Journalism Quarterly* 51, no. 1 (Spring 1974): 111–15.

Wilson, W. Cody, and Herbert I. Abelson. "Experience with and Attitudes Toward Explicit Sexual Materials." *Journal of Social Issues* 29, no. 3 (1973): 19–39.

Winthrop, Henry. "The Sociologist and the Study of the Future." *American Sociologist* 3, no. 2 (May 1968): 136–45.

"Wired Nation Evil or Good? Ferris Wonders." *Variety* 30, no. 2 (November 12, 1980): 47.

Wood, Fred B.; Vary T. Coates; Robert L. Chartrand; and Richard F. Ericson. *Videoconferencing Via Satellite: Opening Congress to the People.* Washington, D.C.: Program of Policy Studies in Science and Technology, 1979.

Woodruff, Jay. "Switch to Satellite Gear to Save Insurer $1 Million." *ComputerWorld*, March 10, 1980. Reprint by American Satellite Corporation.

Xerox. *The Ethernet: A Local Area Network.* Stamford, Conn., 1980.

Zito, Tom. "TV Seers Take the Qube Route." The *Washington Post*, August 24, 1979, pp. B1, B6.

Index